A Student's Guide to Entropy

Striving to explore the subject in as simple a manner as possible, this book helps readers understand the elusive concept of entropy.

Innovative aspects of the book include the construction of statistical entropy from desired properties, the derivation of the entropy of classical systems from purely classical assumptions, and a statistical thermodynamics approach to the ideal Fermi and ideal Bose gases. Derivations are worked through step-by-step and important applications are highlighted in over 20 worked examples. Around 50 end-of-chapter exercises test readers' understanding.

The book also features a glossary giving definitions for all essential terms, a time line showing important developments, and a list of books for further study. It is an ideal supplement to undergraduate courses in physics, engineering, chemistry, and mathematics.

DON S. LEMONS is Professor Emeritus of Physics at Bethel College, where he taught undergraduate physics for 23 years, and a Guest Scientist at Los Alamos National Laboratory. He taught undergraduate physics at Bethel College for 23 years.

A Student's Guide to Entropy

DON S. LEMONS
Bethel College

CAMBRIDGE
UNIVERSITY PRESS

CAMBRIDGE
UNIVERSITY PRESS

University Printing House, Cambridge CB2 8BS, United Kingdom

One Liberty Plaza, 20th Floor, New York, NY 10006, USA

477 Williamstown Road, Port Melbourne, VIC 3207, Australia

314-321, 3rd Floor, Plot 3, Splendor Forum, Jasola District Centre, New Delhi - 110025, India

79 Anson Road, #06-04/06, Singapore 079906

Cambridge University Press is part of the University of Cambridge.

It furthers the University's mission by disseminating knowledge in the pursuit of education, learning and research at the highest international levels of excellence.

www.cambridge.org
Information on this title: www.cambridge.org/9781107653979

First published 2013
Reprinted 2013

A catalogue record for this publication is available from the British Library

Library of Congress Cataloging in Publication data
Lemons, Don S. (Don Stephen), 1949–
A student's guide to entropy / Don S. Lemons.
pages   cm
Includes bibliographical references and index.
ISBN 978-1-107-01156-4 (hardback) – ISBN 978-1-107-65397-9 (paperback)
1. Entropy.   I. Title.
QC318.E57L46   2013
003´.85–dc23   2013009569

ISBN 978-1-107-01156-4 Hardback
ISBN 978-1-107-65397-9 Paperback

Contents

v

Preface

The mathematician John von Neumann once urged the information theorist Claude Shannon to assign the name *entropy* to the measure of uncertainty Shannon had been investigating. After all, a structurally identical measure with the name *entropy* had long been an element of statistical mechanics. Furthermore, "No one really knows what entropy really is, so in a debate you will always have the advantage." Most of us love clever one-liners and allow each other to bend the truth in making them. But von Neumann was wrong about entropy. Many people have understood the concept of entropy since it was first discovered 150 years ago.

Actually, scientists have no choice but to understand entropy because the concept describes an important aspect of reality. We know how to calculate and how to measure the entropy of a physical system. We know how to use entropy to solve problems and to place limits on processes. We understand the role of entropy in thermodynamics and in statistical mechanics. We also understand the parallelism between the entropy of physics and chemistry and the entropy of information theory.

But von Neumann's witticism contains a kernel of truth: entropy is difficult, if not impossible, to visualize. Consider that we are able to invest the concept of the *energy* of a rod of iron with meaning by imagining the rod broken into its smallest parts, atoms of iron, and comparing the energy of an iron atom to that of a macroscopic, massive object attached to a network of springs that model the interactions of the atom with its nearest neighbors. The object's energy is then the sum of its kinetic and potential energies – types of energy that can be studied in elementary physics laboratories. Finally, the energy of the entire system is the sum of the energy of its parts.

These imaginative transitions – first to analyze a whole into its parts, second to compare each part with a familiar object, third to recognize the quantity sought in the familiar object, and finally to recompose the whole out of its parts – fail to shed light on entropy. The difficulty lies in recognizing the entropy in the smallest part. A single, localized molecule or atom has no entropy. Nevertheless, a rod of iron has entropy. Entropy is not a localized phenomenon at which we can point, even in our imaginations, and say: "Look! There is entropy." And, if we insist on trying to understand a subject in ways inconsistent with its nature, we will be disappointed.

This student guide introduces us to the ways in which entropy can be understood. It emphasizes conceptual foundations and exemplary illustrations and distinguishes among different kinds of entropy: thermodynamic entropy, the entropy of classical and quantized statistical systems, and the entropy of information. These entropies differ in the classes of systems to which they apply. But all name the same concept in the sense that they reduce to each other when they can be applied to the same object.

Several features make this text appropriate as a student guide and appropriate for self-study. Never does the text depend upon "It can be shown." Derivations are included and each step of each derivation is made explicit. Mathematical techniques are used and reused and each piece of physics is revisited in different contexts. Worked examples illustrate important applications. Italicized words usually mean, "Note well. Here is a definition." These definitions, often in expanded form, are collected in a Glossary. The answers to all end-of-chapter problems are appended and an Annotated Further Reading list identifies sources and suggests books for further study. Rudolph Clausius, Ludwig Boltzmann, Max Planck, Claude Shannon, and others developed the concept of entropy over a period of 100 years. I have not been reluctant to assign credit where credit is due and to associate people and developments with dates. A Time Line organizes this information. A Formulary contains useful identities and expansions.

The middle chapters of this guide, Chapters 2 to 7, compose an entropy-centered introduction to equilibrium statistical mechanics. Not only is entropy a smooth pathway to major results, including blackbody radiation and the ideal quantum gases, but concentrating on entropy keeps one close to the physics.

The experienced user of thermal and statistical physics texts will notice several original features. One of these is the way statistical entropy is constructed, step-by-step, out of desired properties. This first phase of this construction is accomplished in Section 2.4 and the last in Section 4.5. Another is that the entropy of a classical system is derived from purely classical presuppositions in Section 2.4. Most texts do not bother with constructing a coherent classical statistical mechanics, but instead make do with the semi-classical limit of the corresponding quantum description. A third is that the ideal Fermi gas (Section 6.3) and ideal Bose gas (Sections 7.3 and 7.4) are developed from the properties of their entropy functions via the average energy approximation (Section 6.4) rather than from partition functions and density of states functions. The logic of these constructions and derivations is uncluttered and efficient.

What has been most important to me in writing this guide has been to closely correlate verbal and mathematical formulations and to translate simple concepts into simple mathematics. If the whole is complex, I have always meant the parts to be simple. My goal has been to write simply of the simple and clearly of the complex.

If I have, in some measure, been successful, no small credit is due those who helped me. Ralph Baierlein, in particular, deserves my thanks for thoughtfully reading and rereading the entire text and giving unselfishly of his expert advice. Ralph saved me from several consequential missteps and improved the readability of the text. Ralph also summarized content from German language sources. Anthony Gythiel provided translations from the German. Rick Shanahan also read and commented thoughtfully on the entire text. Dale Allison, Clayton Gearhart, Galen Gisler, Bob Harrington, Carl Helrich, and Bill Peter read and commented on various parts of the text. Hans von Baeyer, Andrew Rex, and Wolfgang Reiter graciously answered my questions. My students in a thermal physics course at Wichita State University constantly reminded me of my purpose. I am indebted to these students, colleagues, and friends.

1

Thermodynamic entropy

1.1 Thermodynamics and entropy

The existence of entropy follows inevitably from the first and second laws of thermodynamics. However, our purpose is not to reproduce this deduction, but rather to focus on the concept of entropy, its meaning and its applications. Entropy is a central concept for many reasons, but its chief function in thermodynamics is to *quantify the irreversibility of a thermodynamic process.* Each term in this phrase deserves elaboration. Here we define *thermodynamics* and *process*; in subsequent sections we take up *irreversibility*. We will also learn how entropy or, more precisely, differences in entropy tell us which processes of an isolated system are possible and which are not.

Thermodynamics is the science of macroscopic objects composed of many parts. The very size and complexity of thermodynamic systems allow us to describe them simply in terms of a mere handful of *equilibrium* or *thermodynamic variables*, for instance, pressure, volume, temperature, mass or mole number, internal energy, and, of course, entropy. Some of these variables are related to others via *equations of state* in ways that differently characterize different kinds of systems, whether gas, liquid, solid, or composed of magnetized parts.

A thermodynamic system undergoes a *thermodynamic process* when the variables that describe the system change from one set of values to another, that is, change from one *thermodynamic state* to another. Thus, we describe a thermodynamic process by specifying an initial state, a final state, and specific conditions under which the change occurs. Specific conditions include: *iso-energetically*, that is, with constant energy as would occur when the system is completely isolated; *adiabatically*, that is, within a thermally insulating boundary; and *quasistatically*, that is, so slowly the system occupies a continuous sequence of thermodynamic states each one of which is characterized by thermodynamic variables. Thermodynamics *per se* says nothing about the rate at which a process unfolds.

The laws of thermodynamics also limit the ways in which a thermodynamic system may proceed from one state to another. The *first law of thermodynamics* is the law of conservation of energy applied to the internal energy of a thermodynamic system. The *internal energy* excludes the energy of the position or motion of the system as a whole. According to the first law there are only two ways to change the internal energy of a given thermodynamic system: (1) heating or cooling the system and (2) working on the system or arranging for the system to work on its surroundings. Quantitatively the first law is

$$\Delta E = Q + W \tag{1.1}$$

where $\Delta E \left[= E_f - E_i \right]$ is the increment in the internal energy E of a thermodynamic system as it transitions from an initial state with energy E_i to a final state with energy E_f. The quantity Q is the energy absorbed by the system when heated, and W is the work done on the system during the transition. These are signed quantities so that when $Q < 0$, the system is cooled rather than heated, and when $W < 0$, work is done *by* rather than *on* the system. (Note that some texts use the symbol W in the opposite way so that when $W < 0$, work is done *on* rather than *by* the system. In this case the first law of thermodynamics is $\Delta E = Q - W$ instead of (1.1).)

When these changes are indefinitely small the first law becomes

$$dE = \delta Q + \delta W. \tag{1.2}$$

Here the different notations, dE versus δQ and δW, emphasize that while E is a variable that describes the state of a system, Q and W are not. The heat quantities, Q and δQ, and their work correlatives, W and δW, merely indicate the amount of energy transferred to or from a system in these ways. (We have no standard way of distinguishing between a differential of a state variable and an equivalently small infinitesimal that is not the differential of a state variable. Here I use d to signal the first as in dE and δ to signal the second and as in δQ and δW. Other texts adopt other solutions to this notational problem.) Thus, a thermodynamic system contains energy but does not, indeed cannot, contain heat or work, as suggested in Figure 1.1.

1.2 Reversible and irreversible processes

All interactions among fundamental particles are *reversible*. The simplest interaction is a collision between two fundamental particles. One particle approaches another; they interact via gravitational, electromagnetic, or nuclear

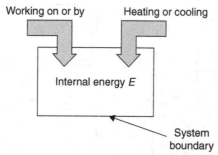

Figure 1.1 The internal energy E of a thermodynamic system can be changed in only two ways: (1) heating or cooling and (2) working.

forces; then they recede from each other. Interestingly, a film of such a collision would make as much physical sense proceeding backwards as proceeding forwards. It may help us to think of a white cue ball striking a colored billiard ball, the latter initially at rest. While some care may be required to set both balls in motion with reversed velocities in such a way that the colored billiard ball strikes the cue ball and both recover their initial states of rest and motion, it can be done. And certainly this reversal can be imagined without supposing the violation of any of Newton's laws of motion. For there is nothing in the laws of motion or in the way fundamental particles interact that prefers one direction of time over the other. We say that interactions among fundamental particles are *time reversible*.

The same cannot be said of the processes of a thermodynamic system composed, as it is, of many parts. Thermodynamic processes are, in fact, typically *irreversible*. A cup of hot coffee sitting on the kitchen table always cools down. We never observe a cup of initially room-temperature coffee extracting energy from the air and heating up. When a hot object and a cold one are in thermal contact, the hot object always cools and the cold object always heats until both reach an intermediate temperature. We never observe these processes occurring in reverse order. A backwards-running video of a thermodynamically irreversible process appears implausible.

Thermodynamic reversibility

There is, however, a special sense in which a thermodynamic process can meaningfully be reversed. If making only an indefinitely small change in the system or its environment reverses the direction of the process, the process is thermodynamically reversible. Sadi Carnot (1796–1832) invented the concept of *thermodynamic reversibility* in order to articulate and prove what is now called *Carnot's theorem*: *The most efficient heat engine is one that operates reversibly.*

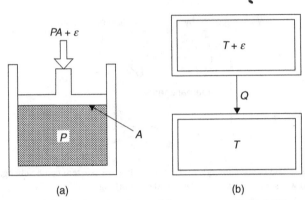

Figure 1.2 (a) A frictionless piston reversibly compresses a fluid. (b) Two systems with an infinitesimal temperature difference heat and cool each other reversibly.

A reversible thermodynamic process must be (1) quasistatic, that is, indefinitely slow, and (2) without friction or dissipation (internal friction). Consider, for instance, the piston compressing the fluid illustrated in Figure 1.2a. The piston quasistatically and frictionlessly compresses the fluid when the force the piston exerts on the fluid, $PA + \varepsilon$, is an indefinitely small amount ε larger than the force PA the fluid of pressure P exerts on the piston head of area A. When, instead, the force exerted by the piston on the fluid is infinitesimally less than PA, that is, $PA - \varepsilon$, the fluid expands quasistatically. Thus, these are thermodynamically reversible changes because their direction can be reversed by an infinitesimal change in either the system or its environment. In similar fashion, the heating or cooling of one system by another is reversible when an infinitesimal temperature difference is maintained between the two as shown in Figure 1.2b. Clearly, thermodynamic reversibility is an ideal that can be approached but never fully realized. All actual thermodynamic processes are irreversible.

To summarize: all fundamental processes are reversible because one can imagine reversing their direction of change without violating a law of physics. On the other hand, all non-idealized thermodynamic processes are irreversible. Thermodynamic processes, in practice, always proceed in one direction.

Loschmidt's paradox

Yet if, as is widely assumed, a thermodynamic system is composed of many fundamental particles and a thermodynamic process is composed of many fundamental interactions, why are not all thermodynamic processes reversible? Johann Loschmidt (1821–1895) asked this question in 1876. We still have no fully satisfactory answer. That many reversible fundamental processes do not necessarily compose a reversible thermodynamic process is known as *Loschmidt's paradox* or the *reversibility paradox*. Our failure to resolve Loschmidt's paradox suggests that the laws governing the interactions of fundamental particles do not form a complete picture of nature and need to be supplemented with additional physics equivalent to the second law of thermodynamics.

Example 1.1 Reversible or irreversible?

Problem: A piston quasistatically compresses a gas enclosed in a chamber. In order to compress the gas the piston must overcome a force of 0.01 N caused by the piston head rubbing against the sides of the piston chamber. Is the work performed on the gas reversible or irreversible?

Solution: This work is performed quasistatically but not without friction. Therefore, the work is irreversible. Alternatively, in order to change the quasistatic compression into a quasistatic expansion the force exerted on the piston must change by the finite amount of 2×0.01 N.

1.3 The second law of thermodynamics

There are a number of logically equivalent versions of the second law. Perhaps not surprisingly, the earliest versions are the most readily grasped. Rudolph Clausius (1822–1888) was the first, in 1850, to identify a version of the second law of thermodynamics. For this purpose Clausius elevated everyday observations about heating and cooling to the status of a general law: *A cooler object never heats a hotter object.*

By *heating (cooling) an object* we mean transferring energy to (from) the object without doing work. And by *cooler object* we mean one with lower temperature and by *hotter object* we mean one with higher temperature. We will discuss *temperature* in more detail later. For now it suffices to consider temperature as the degree of hotness measured on some scale.

Heat reservoirs

Clausius's and other versions of the second law of thermodynamics are most conveniently expressed in terms of *heat reservoirs*. By definition, a heat reservoir maintains the same temperature regardless of how much heating or cooling it experiences. Thus a heat reservoir has infinite heat capacity just as if it were indefinitely large. In the language of heat reservoirs Clausius's second law is: *A process whose only result is to cause one heat reservoir with temperature T_C to lose energy Q through cooling and a hotter heat reservoir with temperature $T_H > T_C$ to gain energy Q through heating never occurs.*

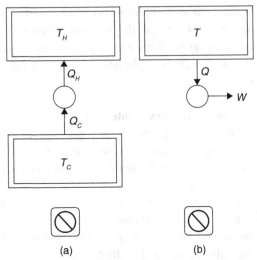

Figure 1.3 (a) A process whose only result is to cause one heat reservoir with temperature T_c to lose energy Q_C through cooling and cause a hotter heat reservoir with temperature $T_H > T_C$ to gain energy Q_H through heating is forbidden by Clausius's version of the second law, in this case $Q_H = Q_C$. (b) A heat engine whose only result is to cause a single heat reservoir to lose energy Q through cooling and perform an equal amount of work $W[=Q]$ is forbidden by Kelvin's version of the second law.

In 1851 William Thomson (1824–1907), later known as Lord Kelvin, formulated a verbally distinct version of the second law. Kelvin's second law concerns the operation of a *heat engine*, that is, a device that uses a temperature difference to produce work. According to Kelvin's second law: *A heat engine whose only result is to cause a single heat reservoir to lose energy Q through cooling and perform an equal amount of work W [= Q] is impossible.*

Given the first law, Kelvin's version of the second law is logically equivalent to Clausius's. The version one favors is a matter of taste and convenience. If one wants a second law in the language of natural phenomena, one adopts Clausius's second law. If one wants a second law in the language of technology, one adopts Kelvin's. We illustrate these impossible processes in terms of which these versions of the second law are framed in Figure 1.3 and label them with a symbol $\oslash$ that proclaims, "forbidden by the second law."

The second law and irreversibility

The second law of thermodynamics can also be expressed in terms of the language of irreversible processes. Recall that an irreversible thermodynamic process cannot be completely reversed. Therefore, another way to say that a particular process is forbidden is to say that the forbidden process is the result of reversing a thermodynamically irreversible process. In these terms Clausius's second law becomes: *A process whose only result is to cool a hotter reservoir and heat a colder one cannot be completely reversed.* Likewise, Kelvin's second law becomes: *A process whose only result is to dissipate work into a single heat reservoir cannot be completely reversed.* Every time our coffee cools and every time we shuffle across the floor we experience or create necessarily irreversible processes. In the next two sections we focus on the first of these, a hotter reservoir heating a colder one, in order to introduce the measure of irreversibility we call entropy and to reformulate the second law in terms of that measure.

1.4 Entropy and irreversibility

Imagine a thermodynamic system that can exist in many different thermodynamic states. We represent each of these states by a point in a multi-dimensional thermodynamic state space. The variables that define the state space will, of course, depend upon the nature of the system. For instance, because only two variables completely determine the state of a *simple fluid*, the state space of a simple fluid is a plane whose points are defined by locating values on two axes: one labeled, say, energy E and the other volume V. Now, choose any two points representing two different states in our thermodynamic state space and a path through the state space that connects these two points as in Figure 1.4. Each point on this path necessarily represents a thermodynamic state and the complete path necessarily represents a quasistatic process that connects the two endpoint states.

Of course, non-quasistatic, for example, turbulent or explosively rapid, processes can also transform a system from one thermodynamic state to another. But non-quasistatic processes are not composed of a continuous sequence of thermodynamic states and cannot be represented by a path in thermodynamic state space.

Therefore, a path in thermodynamic state space necessarily represents a quasistatic process. And quasistatic processes come in only two kinds: reversible processes that proceed without friction or dissipation and irreversible ones that proceed with friction or dissipation. Recall that a completely reversible process is also one for which the system and its environment can, with only infinitesimal adjustments, evolve along a path in state space in either direction, and an irreversible process is one for which the system and its environment would require finite adjustments to proceed along a path in either direction. A piston that very slowly compresses a gas while overcoming friction between

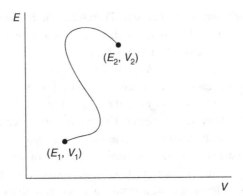

Figure 1.4 Thermodynamic state space, two states, denoted (E_1, V_1) and (E_2, V_2), and a path connecting the two states.

the piston head and the sides of the piston chamber is an example of an irreversible, quasistatic process that can be represented by a path in state space. To summarize: every segment of a path in state space represents a quasistatic process that, in turn, represents either a reversible or an irreversible process.

Entropy difference

We seek a quantitative measure of the irreversibility of a quasistatic process. Rudolph Clausius was, in 1865, able to construct such a measure by discovering a state variable of an isolated system that monotonically changes along a path representing an irreversible process and remains the same along a path representing a reversible process. Thus, two states of an isolated system connected by a reversible path have the same value of this state variable and two states of an isolated system connected by an irreversible path have different values of this state variable. Clausius's discovery of this state variable required an elaborate deduction from the first and second laws of thermodynamics. I will not reproduce that deduction.

Clausius coined the word *entropy* for this state variable – a word parallel in structure, sound, and spelling to the word *energy* – and adopted S for its symbol. Clausius derived the word *entropy* from the Greek root *tropy* meaning "turn" and a prefix *en* meaning *in*. Thus *entropy* literally means "in turn" or "turn in." It seems that Clausius sought a name for a concept that, in part, characterizes how physical systems *turn*, that is, how they *change, evolve,* or *proceed*. Clausius himself interpreted the word *entropy* metaphorically as *transformation content*. We will in time discover more suggestive metaphors for the concept of entropy.

In a thermodynamics based on the first and second laws the only purpose of Clausius's state variable *entropy* is to provide comparisons between two states. For example, if the path connecting states 1 and 2 of an isolated system as illustrated in Figure 1.4 represents an irreversible process, then either $S_2 > S_1$ or $S_1 > S_2$. Alternatively, if the path illustrated represents a reversible process of an isolated system, then $S_2 = S_1$.

The absolute entropy of a single state is much like the absolute energy of a single state. The first and second laws of thermodynamics sanction neither absolute energy nor absolute entropy but assign meaning only to energy and entropy differences. The third law of thermodynamics does, indeed, allow us to assign a meaning to the concept of absolute entropy. But before we explore the third law we determine, in the next several sections, how the entropy difference between two states of an isolated system quantifies the irreversibility of a process connecting those two states.

1.5 Quantifying irreversibility

The key to quantifying the irreversibility of an isolated system's evolution is to require that the entropy increment ΔS of a composite system be additively distributed over its parts. Symbolically stated,

$$\Delta S = \sum_j \left(S_{f,j} - S_{i,j} \right) = \sum_j \Delta S_j \qquad (1.3)$$

where $S_{f,j}$ is the final entropy of the system's jth part, $S_{i,j}$ is the initial entropy of the system's jth part, and $\Delta S_j \left[= S_{f,j} - S_{i,j} \right]$ is the entropy increment of the system's jth part. In this way, and as illustrated in Figure 1.5, the entropy increment of a composite system ΔS is the sum of the entropy increments of its parts.

To proceed further we apply these ideas to a particular process: a heat reservoir with temperature T is heated by absorbing energy Q. If, alternatively, the heat reservoir is cooled, then $Q < 0$. Therefore, if the heat reservoir increments its entropy, that increment ΔS can be a function of only two quantities: the reservoir temperature T and the heat Q where the sign of Q indicates whether energy is absorbed through heating $\left[Q > 0 \right]$ or rejected through cooling $\left[Q < 0 \right]$. [The noun *heat* here refers to the energy absorbed or rejected by heating or cooling.] The equation

$$\Delta S = f(T, Q) \qquad (1.4)$$

$$\Delta S = \Delta S_1 + \Delta S_2 + \Delta S_3 + \Delta S_4 + \Delta S_5$$

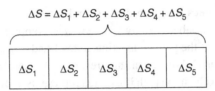

Figure 1.5 The entropy increment ΔS is additive over the parts of a composite system.

states this relation. Note that we have imposed no special restrictions on the system or on this heating or cooling process. In particular the heating or cooling described in Eq. (1.4) may be either reversible or irreversible.

Applying additivity

The additive property (1.3) severely restricts the form of the function $f(T,Q)$. For instance, if we divide a single-temperature T heat reservoir that absorbs heat Q into two identical parts, each still has temperature T and each absorbs half the heat $Q/2$. Since the entropy increment of the composite system is the sum of the entropy increments of its parts: $f(T, Q) = f(T, Q/2) + f(T, Q/2)$ or, equivalently, $f(T, Q) = 2f(T, Q/2)$. If instead of dividing the heat reservoir into two identical parts we divide it into n identical parts, then

$$f(T, Q) = nf\left(T, \frac{Q}{n}\right). \qquad (1.5)$$

There is only one non-trivial solution to (1.5) and that is

$$f(T, Q) = g(T)Q, \qquad (1.6)$$

where $g(T)$ is an as yet undetermined function of the temperature T. (One can derive (1.6) by taking the partial derivative of (1.5) with respect to n, and solving the resulting partial differential equation.) Accordingly, the entropy of a heat reservoir with temperature T changes by

$$\Delta S = g(T)Q \qquad (1.7)$$

when the reservoir absorbs or rejects energy Q through, respectively, cooling or heating another object.

Entropy generation in a two-reservoir, isolated system

Next we apply these ideas to a composite, two-reservoir, isolated system that experiences the following irreversible process: a hotter reservoir directly heats

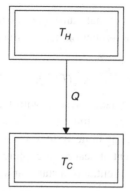

Figure 1.6 A heat reservoir with temperature T_H irreversibly heats a colder reservoir with temperature $T_C < T_H$. In this process energy Q is transferred from the hotter reservoir to the colder reservoir.

a colder reservoir and in so doing transfers energy Q from the hotter reservoir to the colder reservoir as illustrated in Figure 1.6. Since the colder reservoir absorbs energy Q and the hotter reservoir rejects energy Q, their entropy increments are, respectively, $\Delta S_C = g(T_C)Q$ and $\Delta S_H = -g(T_H)Q$. (Here $Q > 0$.) According to these expressions and the additive property (1.3), the entropy increment of this composite two-reservoir system is

$$\Delta S = \left[g(T_C) - g(T_H)\right]Q. \tag{1.8}$$

If the two heat reservoirs of this composite, isolated system have the same temperature $T_H = T_C$, then $\Delta S = 0$. And if their temperatures, T_H and T_C, are different, then heat flows irreversibly from the hotter reservoir to the colder one and $\Delta S \neq 0$. In this way the entropy increment ΔS quantifies the irreversibility of a thermodynamic process of an isolated system. In brief, when the process is reversible $\Delta S = 0$ and when irreversible $\Delta S \neq 0$.

Clausius adopted the convention that the entropy of an isolated system increases when the system evolves irreversibly in the direction allowed by the second law of thermodynamics. Thus, according to Clausius's convention, when $T_H > T_C$ in (1.8), $\Delta S > 0$. This convention means that $g(T_C) - g(T_H) > 0$ and therefore that $g(T)$ must be a non-negative, ever-decreasing function of the temperature T.

Reversible heating and cooling

We can easily overcome the restriction of $\Delta S = g(T)Q$ to heat reservoirs by limiting Q to an indefinitely small quantity δQ. The transfer of an indefinitely small quantity of heat δQ through heating or cooling must, of course, be reversible, since the direction of that transfer can be reversed by an indefinitely small change in the system or its environment. In this case $\Delta S = g(T)Q$ reduces to an expression,

$$dS = g(T)\delta Q, \tag{1.9}$$

that applies to any thermodynamic system with temperature T that absorbs $[\delta Q > 0]$ or rejects $[\delta Q < 0]$ an indefinitely small amount of energy δQ through heating or cooling. After all, any finite system acts like a heat reservoir relative to an indefinitely small amount of heating or cooling. Again, notice the distinction in (1.9) between the differential of a state variable dS and the indefinitely small increment δQ.

Summary

We pause to summarize our progress thus far. The entropy increment ΔS is additive over a system's parts. From this requirement it follows that the entropy increment is $dS [= g(T)\delta Q]$ whenever a thermodynamic system with temperature T absorbs $[\delta Q > 0]$ or rejects $[\delta Q < 0]$ energy δQ through heating or cooling. The function $g(T)$ is a non-negative, ever-decreasing function of T.

According to the work of Rudolph Clausius, the entropy of an isolated system remains constant if and only if it evolves reversibly and increases if and only if it evolves irreversibly. The content of these statements is encapsulated in the entropic version of the second law of thermodynamics: *The entropy of an isolated system cannot decrease.*

Example 1.2   Joule expansion

Problem: A partition separates a container into two parts. One part contains a gas and the other part is evacuated. The partition is quickly removed and the gas fills the whole container in a process called *Joule expansion* as illustrated in Figure 1.7. Does the entropy of the gas increase, decrease, or remain the same, and why?

Solution: According to the entropic version of the second law of thermodynamics, the entropy of an isolated system cannot decrease. The container, partition, and gas constitute an isolated, composite system whose entropy cannot decrease. Since the container does not change and the removal of the partition is a mere spatial rearrangement of a single object, the entropy of these parts of the composite system does not change. Therefore, the entropy of the gas cannot decrease; it can only increase or remain the same. Yet the entropy of an isolated system remains the same only when it undergoes a reversible change.

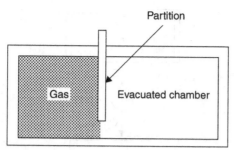

Figure 1.7 Joule expansion of a gas into an evacuated chamber.

And a reversible change occurs only when the direction of the process can be reversed by an infinitesimal change in the system or its environment or, alternatively, when the process is quasistatic and without friction or dissipation. Neither condition is met here. In particular, the quick removal of the partition suggests that this process is not quasistatic. Therefore, the entropy of the gas must increase.

Example 1.3 Humpty Dumpty

Problem: An egg falls from a second floor window, hits the pavement, and breaks. Does the entropy of the egg increase, decrease, or remain the same, and why?

Solution: Again we apply the entropic version of the second law: the entropy of an isolated system cannot decrease. Since the egg hits the pavement we must include the pavement and the egg in the composite isolated system whose entropy cannot decrease. The breaking of the egg is certainly not a reversible change since it cannot be reversed by an infinitesimal change in the egg or in the pavement. Therefore this process is irreversible and, consequently, the entropy of the egg–pavement system increases. It seems reasonable to assume that the pavement does not change much. Therefore, most or all of the entropy increase occurs in the egg.

1.6 The Carnot efficiency and Carnot's theorem

One consequence of the entropic version of the second law, *the entropy of an isolated system cannot decrease*, is Carnot's theorem: *The most efficient heat engine is one that operates reversibly.* To see this observe, that Carnot's simplest heat engine – one that extracts energy Q_H by cooling a hotter reservoir, produces work W, and rejects waste energy Q_C by heating a colder reservoir as

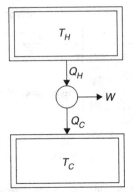

Figure 1.8 The simplest possible heat engine is one that produces work by operating between a single hot and a single cold reservoir. When this simplest heat
engine is reversible it is called a Carnot engine.

illustrated in Figure 1.8 – can be made part of an isolated system by, say, using
the work W to raise a weight. The entropy of this isolated system is

$$\Delta S = -Q_H g(T_H) + Q_C g(T_C), \tag{1.10}$$

where Q_H, Q_C, and W are related, according to the first law, by

$$Q_H = W + Q_C. \tag{1.11}$$

The *efficiency* ε of this engine is, by definition, the engine's benefit, the work
produced W, divided by its cost, the heat it consumes Q_H, that is,

$$\varepsilon = \frac{W}{Q_H}, \tag{1.12}$$

where, given (1.11) and (1.12), we see that $0 \leq \varepsilon \leq 1$. Using (1.11) and (1.12) to
eliminate Q_H and Q_C from the entropy increment (1.10) produces

$$\Delta S = \frac{W g(T_C)}{\varepsilon} \left[1 - \varepsilon - \frac{g(T_H)}{g(T_C)} \right]. \tag{1.13}$$

Therefore, the larger the efficiency ε, the smaller the entropy increment ΔS.
Since the entropy of an isolated system cannot decrease, that is, $\Delta S \geq 0$, the
largest possible efficiency ε is achieved when $\Delta S = 0$, that is, when the engine
operates reversibly. This maximum efficiency of a heat engine operating
between two heat reservoirs,

$$\varepsilon_{\text{Carnot}} = 1 - \frac{g(T_H)}{g(T_C)}, \tag{1.14}$$

is called the *Carnot efficiency* and the reversible version of the simplest possible heat engine is called a *Carnot engine*.

1.7 Absolute or thermodynamic temperature

Neither the entropy increment dS $[= g(T)\delta Q]$ nor the Carnot efficiency $\varepsilon_{\text{Carnot}}$ $[= 1 - g(T_H)/g(T_C)]$ are fully quantified because the non-negative, ever-decreasing, function $g(T)$ is not yet fully specified. William Thomson very cleverly supplied the missing specification in 1848, twenty-four years after Carnot proved Carnot's theorem in 1824 and two years before Rudolph Clausius distinguished and harmonized the first and second laws of thermodynamics in 1850.

Temperature

But first, what is temperature? A thermometer measures temperature, but what is a thermometer? A *thermometer* associates the size of a convenient *thermometric variable*, whether the volume, resistance, color, or some other thermodynamic variable of a system, with a unique number denominated according to a *temperature scale*, whether, for instance, Fahrenheit or Celsius. Consider, for example, the mercury-in-glass thermometer illustrated in Figure 1.9. The length of the mercury column in the small-bore glass tube is the thermometric variable and the way the numbers are arrayed on the side of the glass tube defines the temperature scale.

William Thomson's idea was to use a Carnot engine, that is, a reversible heat engine operating between two heat reservoirs, as a thermometer and its efficiency $\varepsilon_{\text{Carnot}}$ as its thermometric variable. Of course, a Carnot engine is an idealization that can be approached but never realized. After all, a reversible heat engine would operate indefinitely slowly and without friction or dissipation or, alternatively, in such a way that the slightest change in the system or its environment would reverse its operation. Never mind that no one can build a perfectly reversible heat engine and that heat reservoirs are idealized objects! Simply build the best approximation possible and, if necessary, extrapolate the results.

The idea is simple even if its execution is difficult. First, assign a standard temperature $\bar{T}_S$ to some convenient and universally accessible, standard state. One such standard state is the *triple point of water*, that is, the state at which all three phases of water are in equilibrium. Second, measure the efficiency $\varepsilon_{\text{Carnot}}$

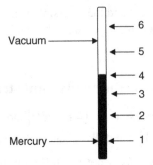

Figure 1.9 Mercury-in-glass thermometer with (arbitrary) temperature scale.

of a reversible heat engine operating between two heat reservoirs, one in equilibrium with the standard state and the other in equilibrium with the system whose temperature $\overline{T}_X$ is to be determined. Third, if $\overline{T}_X < \overline{T}_S$ set

$$\overline{T}_X = \overline{T}_S(1 - \varepsilon_{\text{Carnot}}), \tag{1.15a}$$

and if $\overline{T}_X > \overline{T}_S$ set

$$\overline{T}_X = \frac{\overline{T}_S}{\left(1 - \varepsilon_{\text{Carnot}}\right)}. \tag{1.15b}$$

Temperatures $\overline{T}$ assigned in this way are called *absolute* or *thermodynamic temperatures*, while temperatures T determined by other kinds of thermometers are *empirical temperatures*.

This algorithm allows one to assign an absolute or thermodynamic temperature $\overline{T}$ to any system, in particular to the hot and cold reservoirs between which a reversible heat engine operates. In terms of thermodynamic temperatures assigned according to Eqs. (1.15), the Carnot efficiency of a reversible heat engine operating between two heat reservoirs is given by

$$\varepsilon_{\text{Carnot}} = 1 - \frac{\overline{T}_C}{\overline{T}_H} \tag{1.16}$$

where $\overline{T}_H$ and $\overline{T}_C < \overline{T}_H$ are absolute or thermodynamic temperatures.

Absolute temperature scales

The essential feature of absolute or thermodynamic temperatures is that the ratio of any two temperatures, say $\overline{T}_C/\overline{T}_H$, $\overline{T}_C/\overline{T}_S$, or $\overline{T}_H/\overline{T}_C$, is independent of the scale used to denominate the individual temperatures $\overline{T}_C$, $\overline{T}_H$, or $\overline{T}_S$. For instance, the SI or *Système International* assigns a standard absolute temperature $\overline{T}s = 273.16$ K to the triple point of water. In this way there are, to five places, 100 kelvin degrees between the normal freezing point of water, 273.15 K, and the normal boiling point of water, 373.15 K. The SI unit of absolute temperature, the *kelvin*, and its symbol K are, of course, named after Lord Kelvin (William Thomson). Engineers in the United States and Canada sometimes use the *Rankine scale*, which assigns $\overline{T}s = 491.688°$Ra to the triple point of water. In this way the size of a Rankine degree is identical to the size of a Fahrenheit degree.

In the *Système International* all other kinds of temperatures are defined as linear functions of kelvin temperatures. Translation from one kind of temperature to another is straightforward. For instance, to convert 23 degrees Celsius to kelvin we simply add 273 to 23 and produce, to three-place accuracy, 296 K. And to convert 68 degrees Fahrenheit to Celsius, we subtract 32 from 68 and multiply the result by 5/9 to produce 20 degrees Celsius.

The entropy increment

The function $g(T)$ that defines the entropy increment $dS\left[= g(T)\delta Q\right]$ of a system with temperature T that absorbs heat δQ is constrained by the two expressions for the Carnot efficiency already encountered: (1.14) in terms of empirical temperature and (1.16) in terms of absolute temperature. Thus,

$$1 - \frac{\overline{T}_C}{\overline{T}_H} = 1 - \frac{g(T_H)}{g(T_C)}, \tag{1.17}$$

that is, $\overline{T}_C g(T_C) = \overline{T}_H g(T_H)$. Thus $g(T) = c/\overline{T}$ where c is an arbitrary constant independent of temperature. Choosing $c = 1$ so that $g(T) = 1/\overline{T}$ and that $dS = \delta Q/\overline{T}$ merely fixes the unit of entropy to be that of energy over absolute temperature; for instance, in SI units, joules per kelvin.

While absolute and empirical temperatures are conceptually distinct, maintaining two sets of symbols, $\overline{T}$ and T, serves no practical purpose. For this reason we drop the overbar from $\overline{T}$ and henceforth use the same symbol T for both absolute and empirical temperatures. Context indicates the kind of temperature intended.

The universality and simplicity of absolute temperatures

Absolute temperatures are universal in the sense that their definition is independent of the behavior of any one kind or phase of matter. But their most important advantage is that they reduce all thermodynamic expressions, including all equations of state, to simplest form. For instance, the Carnot efficiency is no longer $\varepsilon_{\text{Carnot}} = 1 - g(T_H)/g(T_C)$ but rather

$$\varepsilon_{\text{Carnot}} = 1 - \frac{T_C}{T_H}. \tag{1.18}$$

And the differential entropy increment $dS = g(T)\delta Q$ becomes

$$dS = \frac{\delta Q}{T}. \tag{1.19}$$

In (1.18) and (1.19) and any other expressions derived from these absolute temperatures, kelvin or Rankine, must be used.

That absolute temperatures simplify all thermodynamic expressions including equations of state follows from the crucial role played by the entropy increment (1.19) in generating equations of state – a topic we address in the next section. For now, reflect on the possibility of expressing the ideal gas pressure equation of state in terms of Fahrenheit temperatures. Certainly, $PV = nR\left[273 + 5(T-32)/9\right]$ is less simple, less beautiful, and less useful than $PV = nRT$!

1.8 Consequences of the second law

Our understanding of the entropy increment $dS = \delta Q/T$ and of the entropic version of the second law of thermodynamics, *the entropy of an isolated system cannot decrease*, has important consequences. One of these is, of course, Clausius's second law. After all, the function $g(T)$ was chosen to be a non-negative, ever-decreasing function of T in order to make the entropic version of the second law equivalent to Clausius's version of the second law.

A second consequence of the entropic second law is Kelvin's second law: An engine whose only result is to cause a single heat reservoir to lose energy Q through cooling and perform an equal amount of work W [$= Q > 0$] is impossible. For the entropy increment $\Delta S = -Q/T$ of this supposed isolated, single-reservoir heat engine is negative and therefore violates the entropic second law. This argument assumes that the work produced by the engine does not change the entropy of the reservoir or its environment. We could, for instance, use this work to raise a weight without changing the entropy of any system. (Note: In this paragraph and throughout this section the symbols Q and W are unsigned, inherently non-negative quantities. Signs indicating whether energy is transferred through heating or cooling and whether work is done on or by the system are inserted into the equations.)

A third consequence is another verbally distinct version of the second law: *A heat engine can be no simpler than one that extracts energy Q_H through cooling a reservoir with temperature T_H, produces work W, and exhausts energy Q_C through heating a reservoir with temperature $T_C < T_H$.* We call this statement Carnot's second law because Sadi Carnot emphasized it in his famous 1824 essay *On the Motive Power of Fire*. Clearly, a heat engine needs both a hot and a cold heat reservoir. For in order that the entropy of the composite, isolated system composed of both reservoirs not decrease, the entropy lost by the hot reservoir $-Q_H/T_H$ must be balanced, or more than balanced, by the entropy gained by the cold reservoir Q_C/T_C, that is, $-Q_H/T_H + Q_C/T_C \geq 0$. Thermodynamics texts rarely, if ever, mention Carnot's second law because Carnot did not know that the assertions he made amounted to a new law of nature. Furthermore, Carnot's analysis employed the caloric theory of heat, according to which in this context $Q_H = Q_C$, rather than the first law of thermodynamics, according to which in this context $Q_H = Q_C + W$. After all, 1824 was at least 20 years before James Joule's precision experiments of the 1840s began to compel acceptance of the first law of thermodynamics.

While we have used the entropic version of the second law, here and in Section 1.6, to derive Carnot's theorem and the Clausius, Kelvin, and Carnot versions of the second law, in actual fact, all four results were articulated before Rudolph Clausius formulated the concept of entropy in 1865 – Carnot's second law and theorem in 1824, Clausius's second law in 1850, and Kelvin's second law in 1851. Any one of these three formulations of the second law, along with the first law, is a sufficient foundation for classical thermodynamics. In particular, from any one it is possible to prove Carnot's theorem and also, what earlier we merely asserted: A state variable *entropy S* exists whose increment ΔS is additive over the parts of a composite system.

Entropy and stability

Finally, we mention a consequence of the entropic version of the second law of thermodynamics of great practical importance. According to the entropic second law, *the entropy of an isolated system cannot decrease.* Therefore, if an isolated system is constrained from increasing its entropy, say, for instance, by adiabatic walls that separate its hot and cold parts, the system's entropy cannot change, and the system is said to be *thermodynamically stable*. Alternatively, if the entropy of an isolated system can increase because, for instance, its hotter parts can heat its cooler parts, the system is said to be thermodynamically unstable. Eventually, when the different parts of this system achieve thermal equilibrium, a thermodynamically unstable system returns to stability.

However, the entropy is not always a convenient means of determining thermodynamic stability because we often observe or experiment with systems that are not isolated. Instead, quite often the temperature and pressure of a thermodynamic system are maintained by thermal and mechanical contact with an environment. Consider, for instance a liquid in a beaker open to the environment. In this case, the crucial question is not whether the entropy of such a system can *increase* but rather whether its Gibbs free energy

$$G = E - TS + PV \qquad (1.20)$$

can *decrease*. If its Gibbs free energy G can decrease, the system is unstable; if its Gibbs free energy G cannot decrease, the system is stable. This relation (1.19) of the Gibbs free energy G and the entropy S suggests that a minimum in the Gibbs free energy correlates with a maximum in the entropy. Both conditions signal stability, but they apply in different circumstances: a maximum of the entropy applies to an isolated system and a minimum of the Gibbs free energy applies to a system whose temperature and pressure are held constant. We will exploit a stability condition based on the Gibbs free energy in Chapter 7.

Example 1.4 Automobile engine efficiency

Problem: A car engine operates at 37% efficiency but this efficiency is only half of its Carnot efficiency. Suppose the effective cold reservoir of a car engine is the outside air with temperature 300 K and the burning of the fuel–air mixture in the piston combustion chamber creates and maintains the hot reservoir. What is the absolute or thermodynamic temperature of this hot reservoir?

Solution: This solution applies the Carnot efficiency $\varepsilon_{\text{Carnot}} = 1 - T_C/T_H$. From the problem statement we know that $\varepsilon_{\text{Carnot}} = 2 \times 0.37$ and $T_C = 300$ K. Solving for T_H we find $T_H = T_C/(1 - \varepsilon_{\text{Carnot}})$. Substituting these values into the right-hand side produces $T_H = 1154$ K.

Example 1.5 The entropy generator

Problem: The entropy generator consists of a weight of mass m that as it falls turns a paddle wheel immersed in a viscous liquid. See Figure 1.10. The liquid is in thermal contact with a heat reservoir with temperature T. Assume the weight falls quasistatically. How much does the entropy of the viscous liquid and of the reservoir change?

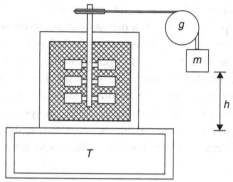

Figure 1.10 Entropy generator.

Solution: As the mass m falls quasistatically through a distance h work mgh is done on the viscous liquid. This would lead to an increase in the internal energy of the liquid and thus to an increase in its temperature were it not that the liquid remains in thermal equilibrium with the reservoir. Therefore, as the weight falls a distance h the liquid transfers energy mgh in heat to the reservoir and the reservoir entropy increases by mgh/T. The entropy of the liquid remains constant. If we consider the liquid and the reservoir as a single composite system, the entropy generator demonstrates that when work W is completely dissipated in a heat reservior of temperature T, the entropy of that system increases by W/T.

1.9 Equations of state

If we use $dS = \delta Q/T$ to eliminate δQ from $dE = \delta Q + \delta W$, we produce a result,

$$dE = T\,dS + \delta W, \tag{1.21}$$

that suggests a relation exists between the internal energy E, the temperature T, and the entropy S of a thermodynamic system. Such relations, or equations of state, do exist and describe what can be known about a thermodynamic system. Equation (1.21) requires that what can be known in this way is consistent with the first and second laws of thermodynamics. The entropy state variable S plays a key role in this linkage. In particular, *equations of state are consistent with the first and second laws of thermodynamics if and only if they can be derived from the entropy S via Eq. (1.21).*

The equation $dE = TdS + \delta W$ is not quite sufficient to demonstrate this fact. Recall Example 1.1 in which a piston head compresses a gas enclosed in a chamber. The work of compression δW included not only the work on the gas but also the work required to overcome the friction between piston head and chamber wall. Only when the work δW is performed reversibly, that is, quasistatically and without friction or dissipation is δW completely described by the thermodynamic variables of the gas system. Similarly, only when heat δQ is transferred reversibly between two systems, that is, quasistatically and therefore reversibly, do these two systems have the same temperature.

Reversible work

When the work done on a system is reversible, $\delta W = \delta W_{rev}$, and (1.21) becomes

$$dE = T\, dS + \delta W_{rev}. \tag{1.22}$$

However, there are as many ways of doing reversible work δW_{rev} as there are kinds of systems. When the system is a *simple fluid* with scalar pressure P and volume V,

$$\delta W_{rev} = -P\, dV. \tag{1.23}$$

The negative sign ensures that positive, reversible work $\delta W_{rev} > 0$ on the fluid decreases the fluid volume by dV. Simple fluids always push; they never pull. When the system is a surface with area A and surface tension σ,

$$\delta W_{rev} = \sigma\, dA. \tag{1.24}$$

Positive, reversible work $\delta W_{rev} > 0$ on the surface increases the surface area by dA. Surfaces of this kind always pull; they never push. When the system is a paramagnet, the currents that create the externally applied magnetic field B_o in which the paramagnet is immersed do reversible work

$$dW_{rev} = \mu_o B_o\, dM \tag{1.25}$$

by increasing the magnetization M of the paramagnet.

In each of these examples the reversible work done on a system δW_{rev} is a product of an intensive variable (pressure P, surface tension σ, or externally applied magnetic field B_o) times the differential of an extensive variable (volume V, area A, or net magnetization M). This factorization of δW_{rev} into an *intensive variable* times an increment of an *extensive variable* can always be accomplished. An extensive variable is one, for instance, that doubles when a system is doubled in size, while an intensive variable remains the same as a system is scaled up or down.

Fluid systems

Fluid systems are especially important. Gases, liquids, and even solids can be treated as fluids in varying degrees of approximation. Combining $dE = TdS + \delta W_{rev}$ and $\delta W_{rev} = -PdV$ produces the *fundamental constraint*

$$dE = TdS - PdV \qquad (1.26)$$

among the variables of a simple fluid, E, T, S, P, and V, imposed by the first and second laws of thermodynamics. We often prefer writing the fundamental constraint (1.26) in a form,

$$dS = \frac{1}{T}dE + \frac{P}{T}dV, \qquad (1.27)$$

that emphasizes the role of entropy. For if the entropy is a function of the independent variables E and V so that $S = S(E,V)$, then its multivariate differential is

$$dS = \left(\frac{\partial S}{\partial E}\right)_V dE + \left(\frac{\partial S}{\partial V}\right)_E dV. \qquad (1.28)$$

(Here $(\partial S/\partial E)_V$ means the derivative of S with respect to E while holding V constant. This use of subscripts in this way, common in thermodynamics texts, has the advantage of reminding us, in this case, that E and V are the two independent variables. After all, other pairs of variables can be designated as independent.) Equations (1.27) and (1.28) together imply that

$$\left(\frac{\partial S}{\partial E}\right)_V = \frac{1}{T} \qquad (1.29)$$

and

$$\left(\frac{\partial S}{\partial V}\right)_E = \frac{P}{T}. \qquad (1.30)$$

Furthermore, the equality of mixed partials, that is, $\partial^2 S/\partial V \partial E = \partial^2 S/\partial E \partial V$, means that for any fluid

$$\left[\frac{\partial}{\partial V}\left(\frac{1}{T}\right)\right]_E = \left[\frac{\partial}{\partial E}\left(\frac{P}{T}\right)\right]_V, \qquad (1.31)$$

where the expressions $1/T$ and P/T are themselves functions of E and V through (1.29) and (1.30).

An example

The equations (1.29) and (1.30) show us how, by taking derivatives of the entropy function $S(E,V)$, to derive fluid equations of state that observe the first and second laws of thermodynamics. For instance, the entropy function for n moles of ideal gas is

$$S(E,V) = nR \ln (\mathrm{VE}^{C_V/nR}) + c \tag{1.32}$$

where R is the gas constant, C_V is the, presumed constant, heat capacity at constant volume, and c is a constant independent of all thermodynamic state variables. Taking the partial derivatives of $S(E,V)$ we find that $(\partial S/\partial E)_V = C_V/E$ and $(\partial S/\partial V)_E = nR/V$. Using these in (1.29) and (1.30) produces the two ideal gas equations of state:

$$PV = nRT \tag{1.33}$$

and

$$E = C_V T. \tag{1.34}$$

In this way, the entropy function (1.32) encapsulates the physics of an ideal gas. Consequently, the equations of state, (1.33) and (1.34), satisfy the requirement (1.30) on mixed partial derivatives.

Working backwards

But how, in the first place, are entropy functions produced? One way to produce an entropy function $S(E,V)$ is to work backwards from known equations of state by integrating $(\partial S/\partial E)_T = 1/T$ and $(\partial S/\partial V)_E = P/T$. Suppose, for instance, we start with the equations of state for *equilibrium* or *blackbody radiation*

$$P = \frac{E}{3V} \tag{1.35}$$

and

$$E = aVT^4, \tag{1.36}$$

where a is a universal constant called the *radiation constant*. Blackbody radiation can be considered a simple fluid because it occupies a volume V and exerts a scalar pressure P. Solving (1.35) and (1.36) for P/T and T in terms of E and

V allows us to eliminate the former from the right hand sides of $(\partial S/\partial E)_V = 1/T$
and $(\partial S/\partial V)_E = P/T$. In this way we generate

$$\left(\frac{\partial S}{\partial E}\right)_V = \frac{a^{1/4}V^{1/4}}{E^{1/4}} \tag{1.37}$$

and

$$\left(\frac{\partial S}{\partial V}\right)_E = \frac{a^{1/4}E^{3/4}}{3V^{3/4}} \tag{1.38}$$

Integrating (1.37) in the variable E while holding V constant produces

$$\int \left(\frac{\partial S}{\partial E}\right)_V dE = a^{1/4}V^{1/4} \int \frac{dE}{E^{1/4}} \tag{1.39}$$

and, therefore,

$$S(E,V) = \frac{4a^{1/4}V^{1/4}E^{3/4}}{3} + f(V), \tag{1.40}$$

where in this case the "integration constant" $f(V)$ is an undetermined function
of V. Similarly integrating (1.38) in V while holding E constant produces

$$S(E,V) = \frac{4a^{1/4}V^{1/4}E^{3/4}}{3} + g(E), \tag{1.41}$$

where $g(E)$ is another undetermined function. We determine $f(V)$ and $g(E)$ by
requiring that the two expressions for the entropy, (1.40) and (1.41), lead to the
same result. Thus we set the right-hand sides of (1.40) and (1.41) equal to each
other and find that $f(V) = g(E)$. Therefore, the entropy function of blackbody
radiation is

$$S(E,V) = \frac{4a^{1/4}V^{1/4}E^{3/4}}{3} + b \tag{1.42}$$

where b is a constant independent of both E and V.

A summary

We summarize. The entropy is an additive state variable that is a function of
the energy E and the other extensive variables of the system. The partial derivatives of the entropy function in terms of its extensive variables generate the
equations of state of the system. In particular, $(\partial S/\partial E)_V = 1/T$. Alternatively,
we can produce the entropy function of a system from its equations of state.

Entropy could also be expressed in terms of a mixture of extensive variables and intensive variables. However, expression exclusively in terms of extensive variables E, V, and mole number n, or particle number N, is for a simple fluid the most fundamental. For this reason we will emphasize the functions $S(E,V)$ and $S(E,V,n)$ or $S(E,V,N)$. In other texts expect to see entropy expressed in terms of a great variety of variables.

The next section introduces a requirement on the entropy function imposed by the third law of thermodynamics.

Example 1.6 Modifying the ideal gas

Problem: Suppose we modify the ideal gas pressure equation of state by intro-ducing attraction between molecules that reduces the pressure exerted by the fluid. We use the parameter a where a is independent of any thermodynamic variable to mediate this attraction. Our modification is

$$P = \frac{nRT}{V} - \frac{an^2}{V^2}.$$

In what way must the energy equation of state, $E = C_V T$, be modified in order to make this pair of equations of state consistent with the first and second laws of thermodynamics?

Solution: There is no infallible algorithm for answering this question. We start by guessing that the modification takes the form

$$E = C_V T + f(V),$$

where we insist that $f(V)$ is chosen so that the first and second laws of thermo-dynamics are satisfied and that C_V remains a constant. In particular, we solve the two displayed equations of state for $1/T$ and P/T in terms of the independent variables E and V and find that $1/T = C_V/[E - f(V)]$ and $P/T = nR/V - aC_V n^2/\{V^2[E - f(V)]\}$. Substituting these results into the requirement (1.31) on mixed partials produces

$$\left\{ \frac{\partial}{\partial V}\left[\frac{C_V}{E - f(V)} \right] \right\}_E = \left\{ \frac{\partial}{\partial E}\left[\frac{nR}{V} - \frac{aC_V n^2}{V^2[E - f(V)]} \right] \right\}_V .$$

Completing these partial derivatives we find that $f(V)$ must satisfy the differ-ential equation

$$\frac{d}{dV}f(V) = \frac{an^2}{V^2}.$$

A solution of this equation is $f(V) = -an^2/V$. Therefore, the modified energy equation of state is, up to an arbitrary constant, given by $E = C_vT - an^2/V$.

1.10 The third law of thermodynamics

Energy and entropy are thermodynamically parallel concepts. Each is an additive state variable. And the existence of each is implied by the laws of thermodynamics: energy by the first law and entropy by the first and second laws. Furthermore, these laws identify ways of measuring their increments. Recall that $dE = \delta Q + \delta W$ and $dS = \delta Q/T$. Thus, one can measure the energy and entropy increments ΔE and ΔS by carefully controlling and measuring the energy transferred to or from a system by heating and cooling or by having work done on or by a system.

Yet energy and entropy play different roles within the science of thermodynamics. The energy of an isolated system remains constant, while the entropy of an isolated system cannot decrease. The third law of thermodynamics identifies another behavior of entropy that has no energetic parallel.

The empirically supported, physical content of the third law is: *The entropy of a system approaches a definite and finite value that is independent of the other thermodynamic variables as its thermodynamic temperature approaches absolute zero.* Max Planck (1858–1948) added to this empirical content an extremely useful convention: *The entropy $S \to 0$ as the thermodynamic temperature $T \to 0$.* A careful analysis of Planck's convention reveals that it actually consists of two statements: (1) the entropy of all systems approach the same constant as $T \to 0$, and (2) the entropy approached is $S = 0$. The empirical content of the third law plus Planck's convention sum to the following version of the third law: *The entropy of every thermodynamic system approaches zero as its thermodynamic temperature approaches absolute zero.* Symbolically, $S \to 0$ as $T \to 0$.

The physical content of the third law constrains the low temperature behavior of thermodynamic systems. Furthermore, Planck's convention makes it possible to define the absolute entropy of a system and so to produce tables of absolute entropies. The third law also has an interesting quantum statistical interpretation – a subject we take up in Chapter 4.

An illustration

Here we simply observe whether or not the entropy of various model systems obeys the third law of thermodynamics, that is, whether or not $S \to 0$ as $T \to 0$. For example, the entropy of an ideal gas is, in its most general form,

$$S(E,V) = nR\ln V + C_V \ln E + c, \qquad (1.43)$$

where c stands for a constant independent of the any thermodynamic variable. We use the energy equation of state $E = C_V T$ to eliminate E from the entropy function (1.43) in favor of T. Doing so produces

$$S(T,V) = nR\ln V + C_V \ln T + c, \qquad (1.44)$$

where the constant c has been redefined. As $T \to 0$, this function diverges to negative infinity. Thus, there can be no, conventionally chosen, finite value of the constant c that makes $S \to 0$ as $T \to 0$. Therefore, the entropy of an ideal gas violates the third law of thermodynamics.

However, the ideal gas and other non-third-law-compliant models remain useful in high temperature regimes. If a model system violates the first or second laws of thermodynamics, it is useless and we discard it. But if a model violates the third law of thermodynamics, we simply recognize its limited validity. For this reason the third law of thermodynamics is neither as foundational nor as consequential as the first and second laws. Even so, the empirical content of the third law is invariably observed and cannot be derived from the other laws of thermodynamics.

Example 1.7 The third law and blackbody radiation

Problem: Determine whether or not the blackbody radiation equations of state obey the third law of thermodynamics.

Solution: The blackbody radiation equations of state are $P = E/3V$ and $E = aT^4$. According to Eq. (1.42) these imply the entropy function $S(E,V) = b + 4a^{1/4}V^{1/4}E^{3/4}/3$ where b is a constant independent of E and V. Using the energy equation of state $E = aT^4$ to eliminate E in $S(E,V)$ in favor of the temperature T produces $S(T,V) = b + 4aV^{1/4}T/3$. Therefore, $S \to b$ as $T \to 0$ and the constant b can be chosen to be zero. The blackbody radiation equations of state obey the third law.

Problems

1.1 Compressing a fluid

A weight falls on a piston and irreversibly and adiabatically compresses the fluid system contained in the piston chamber (Figure 1.11). Does the entropy of the fluid system increase, decrease, or remain the same and why?

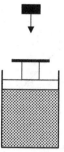

Figure 1.11 A weight falls on piston that irreversibly compresses a gas. See Problem 1.1.

1.2 Derivation

Derive Eqs. (1.6) from (1.5). (See hint in text.)

1.3 Heat capacity

An object with constant heat capacity $C\left[=\delta Q/dT\right]$ absorbs heat and its temperature rises from T_i to T_f. What is its entropy increase?

1.4 Isothermal compression of ideal gas

Consider an ideal gas whose pressure equation of state is $PV = nRT$ and whose internal energy $E = E(T)$ is an unknown function of temperature. What is the increase in its entropy $\Delta S\ [= S_f - S_i]$ as the gas volume is changed from an initial value V_i to a final value V_f while keeping the temperature T constant? (Hint: Integrate $dS = dE/T + P\,dV/T$.)

1.5 Entropy increment

An ideal gas whose equations of state are $PV = nRT$ and $E = 5nRT/2$ evolves quasistatically from state A to state B to state C as shown in Figure 1.12. Find the net entropy increment $\Delta S = S_C - S_A$ per mole of ideal gas. (Hint: Use the graph to integrate $dS = dE/T + PdV/T$.)

1.6 Valid and invalid equations of state

In the following fluid equations of state the symbols a and b stand for positive constants that are independent of all thermodynamic variables.

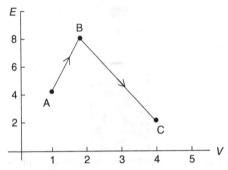

Figure 1.12 The quasistatic path taken by an ideal gas. Used in Problem 1.5.

(1) $PV = aT$ and $E = b\sqrt{T}$
(2) $P = E/V$ and $E = bT$
(3) $P = aT$ and $E = bTV$
(4) $P = aT \sin(E/V)$ and $E = -aT \cos(bEV)$
(5) $P = aE$ and $E = bT$

(a) Identify which two of these five pairs of equations of state do not obey the first and second laws by identifying which do not observe the equality of mixed partials $\dfrac{\partial^2 S}{\partial V \partial E} = \dfrac{\partial^2 S}{\partial E \partial V}$.

(b) Derive the entropy function of the three pairs of equations of state that do obey the first and second laws.

1.7 Entropy function

An entropy function of a simple fluid takes the form $S(E,V) = aEV^2$ where a is a positive constant. Derive the two equations of state of this hypothetical simple fluid.

1.8 Room-temperature solid

One version of the entropy function of a room-temperature solid is

$$S(E,V) = \frac{\alpha_{Po} V}{\kappa_{To}} + C_V \ln\left\{ \frac{E}{E_o} - \frac{(V - V_o)^2}{2\kappa_{To} V_o E_o} \right\},$$

where α_{Po}, κ_{To}, C_V, E_o, and V_o are positive constants whose values characterize the system. (a) Find the energy E and pressure P of the room-temperature solid as functions of V and T. (b) Does this system observe the third law of thermo-dynamics? Why or why not?

1.9 Valid and invalid entropy functions

Some of the following entropy functions $S(E,V)$ for a simple fluid do not result in a positive thermodynamic temperature, some do not observe the third law of thermodynamics, some violate both requirements, and some violate neither requirement. Identify which of the following entropy functions are not valid and indicate the reason or reasons for their invalidity. The constants $a > 0$ and $b > 0$ and are independent of all thermodynamic variables. Also $P > 0$ and $V > 0$.

(1) $S(E,V) = b \ln(EV)$
(2) $S(E,V) = b \ln(EV^2)$
(3) $S(E,V) = b \ln(V/E^2)$
(4) $S(E,V) = b \ln(E/V)$
(5) $S(E,V) = a(EV)$
(6) $S(E,V) = a \exp(-EV)$
(7) $S(E,V) = aV\sqrt{E}$

2

Statistical entropy

2.1 Boltzmann and atoms

The thermodynamic view of a physical system is the "black box" view. We monitor the input and output of a black box and measure its superficial characteristics with the human-sized instruments available to us: pressure gauges, thermometers, and meter sticks. The laws of thermodynamics govern the relations among these measurements. For instance, the zeroth law of thermodynamics requires that two black boxes each in thermal equilibrium with a third are in thermal equilibrium with each other, the first law that the energy of an isolated black box can never change, and the second law that the entropy of an isolated black box can never decrease. According to these laws and these measurements each black box has an entropy function $S(E, V, \ldots)$ whose dependence on a small set of variables encapsulates all that can be known of the black box system.

But we are not satisfied with black boxes – especially when they work well! We want to look inside a black box and see what makes it work. Yet when we first look into the black box of a thermodynamic system we see even more thermodynamic systems. A building, for instance, is a thermodynamic system. But so also is each room in the building, each cabinet in each room, and each drawer in each cabinet. But actual thermodynamic systems cannot be subdivided indefinitely. At some point the concepts and methods of thermodynamics cease to apply. Eventually the subdivisions of a thermodynamic system cease to be smaller thermodynamic systems and instead become groups of atoms and molecules.

The Austrian physicist Ludwig Boltzmann (1844–1906) thought in terms of atoms and molecules at a time when their reality could still be questioned. The distinction between atoms and molecules was not always important to Boltzmann, but he insisted on their reality. In particular, Boltzmann related the entropy of a thermodynamic system to the possible dispositions of its atoms and re-interpreted the second law of thermodynamics in terms of the number of its possible atomic arrangements. In doing so Boltzmann helped create the statistical approach to thermodynamics. But these achievements cost Boltzmann much effort and were in his day strongly resisted.

While the concept of an atom as an indivisible, ultimate particle originates in the fifth century BCE with the Greek thinkers Leucippus and Democritus, the first empirical evidence of their existence came much later, around 1800, as a result of Joseph Proust's (1754–1826) and John Dalton's (1766–1844) interpretation of the way chemical elements combine into compounds. But neither Proust's and Dalton's interpretation nor Boltzmann's speculations compelled a belief in atoms.

Ernst Mach (1838–1916), an influential German-speaking contemporary and some-time colleague of Boltzmann, flatly rejected the existence of atoms. According to Mach and those who adopted his philosophy of science, atoms add little or nothing to our understanding of the world. The goal of science is to correlate directly observable quantities as precisely and as economically as possible. In this sense thermodynamics is the perfect science because thermodynamics achieves this goal without invoking the existence of invisible entities. Given what was known and not known at the time, Boltzmann found Mach's criticism difficult to counter.

Boltzmann's position was not helped by his inadequacies as a writer. Although he was an outstanding scientist, Boltzmann's books and papers are neither concise nor clear. Rather, they probe intuitively and in the process mask important assumptions. Thus Boltzmann invited criticism. And that his life's work was challenged rather than affirmed troubled him. At the age of 62 on September 5, 1906, beset by oncoming blindness and periods of depression, Boltzmann took his own life. Boltzmann was unaware at the time that his long struggle to introduce atoms into physics had been vindicated. For Albert Einstein (1879–1955) in 1905 had shown in a paper on Brownian motion that the random motions of small but visible particles suspended in water were a direct consequence of the random motions of yet smaller invisible atoms and molecules.

While Boltzmann's path of discovery took him through his celebrated H-theorem, our path exploits the insights of those who since Boltzmann's death have reflected on the meaning of his work. Even so, in this chapter we remain faithful to ideas that build directly upon Boltzmann's contributions.

2.2 Microstates and macrostates

When Boltzmann visualized the interior of a thermodynamic system he saw not only atoms and molecules but also their possible arrangements. But what is

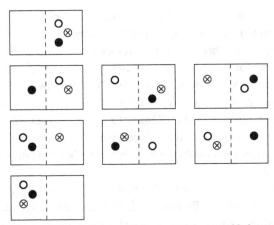

Figure 2.1 The possible microstates of a three-particle, two-sided system.

an *arrangement* or as we now say a *microstate*? (The German word Boltzmann used is sometimes translated as *complexion*.)

Consider a simple system that contains only three particles. We represent these particles, in Figure 2.1, with the symbols: ●, ○, and ⊗. They could be atoms of three different elements or could be identical particles that are in some other way distinguishable. The system is divided into two equal-volume parts, left and right sides, between which the particles may move freely. While simple, this system is just complex enough to illustrate the distinction between a system, its macrostates, and their microstates.

A *microstate* is identified by a detailed particle-level description of the system. Because each of the three particles of this system may be in one of its two sides, this system may realize $2^3 = 8$ different microstates. Each one of these 8 microstates is illustrated in Figure 2.1.

A *macrostate* consists of a set of microstates that can be described with a relatively small set of variables. In more realistic contexts a macrostate is a thermodynamic or equilibrium state and the variables in terms of which it is described are the thermodynamic variables $E, V, N, \ldots$. Consider the macrostate of the three-particle, two-sided system in which any two particles are in the right side of the volume and the remaining particle is in the left. We see from Figure 2.1 that this macrostate consists of just three microstates – those illustrated in the second row of the figure. This number of microstates can also be calculated by choosing two particles (for the purpose of putting them in the right half) from among the three available. We may choose the first particle in three ways and the second in two ways from among the remaining particles for a total of 6 or $3! = 3 \times 2 \times 1$ ways. Since the order in which the two particles are placed into the right side is irrelevant for the purpose of counting microstates, we divide 6 by the number of ways of ordering two particles, that is, by 2 or 2!=2 × 1. In this way we find the number of microstates in the macrostate in which two of three particles occupy the rightmost chamber to be 3 or 3!/(2!1!).

Generalizations

This simple example can be generalized in several ways. If the system contains N particles instead of 3, and still has two different sides that can be occupied by each particle, the binomial coefficient $N!/\left[n!(N-n)!\right]$, that is, "N choose n," identifies the number of ways in which it is possible to distribute N distinct items between the two sides, one with n particles and the other with $N-n$ particles.

Alternatively, if a particle on the left has energy ε and a particle on the right has energy 2ε and we constrain the total energy E of the three-particle system to be 4ε, then the three microstates illustrated in the third row of Figure 2.1 compose this macrostate.

Distinguishable classical particles

In describing these microstates and macrostates we have distinguished each particle from all the others – graphically so in Figure 2.1. Classical particles can always be distinguished from one another. This is not because classical particles necessarily have individually distinguishing features, but rather because their different trajectories through space and time remain distinct.

Imagine, for instance, that you display two apparently identical steel ball bearings to a friend and have your friend cover, and then, after a few seconds, uncover their eyes. While your friend does not know if you have exchanged the two ball bearings, you do know because you have kept your eyes open and never lost track of their different trajectories. Ball bearings are models of classical particles and identically machined ball bearings are models of identical classical particles.

It would have been natural for Maxwell, Boltzmann, and other nineteenth-century physicists to suppose that the invisible, sub-microscopic world of atoms and molecules is simply a smaller version of the visible, macroscopic world. Indeed, Boltzmann argued that the ultimate particles, atoms and molecules, are always distinguishable one from another. In this chapter and the next we adopt this classical assumption that identical particles, for instance, populating a container of helium gas, are just as distinguishable from each other as ball bearings and for the same reason, namely, that we can, at least according to the principles of classical physics, follow their trajectories with indefinite precision.

2.3 Fundamental postulate

According to the *fundamental postulate* of statistical mechanics: *the micro-states of an isolated thermodynamic system are equally probable.* Since the simple three-particle, two-sided system illustrated in Figure 2.1 has 8 microstates consistent with its description and each, according to the fundamental postulate, is equally probable, the probability that this system realizes any one of these 8 microstates is 1/8. Furthermore, because the macrostate with exactly two particles on its right side is realized in 3 out of 8 equally probable microstates, the probability of this macrostate is 3/8.

Multiplicity

The assumed equal probability of the microstates of an isolated system is the material out of which we compose the probability of a macrostate. According to the fundamental postulate, the probability of a macrostate is proportional to the number of microstates contained in a macrostate. Boltzmann called the number of microstates that compose a macrostate its *Permutibilität*. We use the English word *multiplicity* and denote multiplicity with the symbol Ω.

Limitations and counter-examples

The fundamental postulate does not apply to all isolated systems but only to those whose microstates are defined in such a way that their equal probability is plausible. Consider, for instance, a single particle contained within a two-dimensional, square box as shown in Figure 2.2. The particle reflects smoothly and elastically from the sides of the box in such a way that it periodically retraces its path. When we superimpose a uniform grid that identifies the spatial microstates available to this two-dimensional, single-particle system, we see that its microstates are not all equally probable. In fact, most of the system microstates are never occupied. Furthermore, adding more particles to this box may not help realize the fundamental postulate unless the particles are allowed to interact and even then special initial conditions may exist for which not all the possible microstates are realized with equal probability.

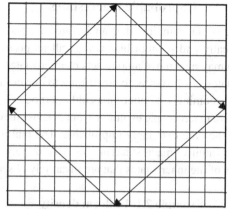

Figure 2.2 A particle reflects smoothly and elastically from the walls of a two-dimensional box and periodically retraces its trajectory. The superimposed grid identifies the microstates available to this one-particle system.

Counter-examples to the fundamental postulate involve special initial conditions of a system with only a few particles. In contrast, the initial conditions of a typical thermodynamic system are unknown and these systems, such as a cubic centimeter of air, contain several times 10^{19} particles. Their structure and evolution is so complex one cannot keep track of the many microstates they occupy. We adopt the fundamental postulate as a way of quantifying our partial knowledge of the state of such a system. In particular, we adopt the fundamental postulate in order to avoid biasing the assignment of probabilities when we have no reason to prefer one microstate over another. Still, the fundamental postulate has not been derived from yet more fundamental principles and cannot be directly tested. Instead, one adopts the postulate of equiprobable microstates as a plausible hypothesis from which one derives consequences that can be tested.

Thermodynamic systems

We call a system to which the fundamental postulate applies a *thermodynamic system*. In general, thermodynamic systems are equilibrium or steady-state systems in the sense that they can be described in term of thermodynamic variables that do not change with time as long as the environment of the system does not change in time. Thermodynamic systems can always be described with a few thermodynamic variables and their macrostates can always be characterized by their multiplicity.

Improbable macrostates

The fundamental postulate implies that an isolated thermodynamic system moves probabilistically and without bias among all its possible microstates. But if so, a system may also move among sets of microstates that correspond to unusual macrostates, for instance, to air rushing to one corner of a room or to a cup of hot tea becoming yet hotter on a cold afternoon. We never observe such counter-intuitive macroscopic evolutions, not because they are impossible but rather because they are extremely improbable. The multiplicities of such unobserved counter-intuitive macrostates are overwhelmingly small compared to the multiplicities of the macrostates that are invariably observed.

Example 2.1 Uniform density

Problem: Each particle in an N-particle gas may move freely between right and left chambers of equal volume. Because these chambers have equal volumes, the probability that any one particle resides in any one chamber is 1 out of 2 or 1/2. In this way, each distinct division of the N distinguishable particles between the two chambers models a single microstate each one of which is equiprobable. Show that the most probable macrostate is one in which the two chambers contain equal numbers of particles.

Solution: The probability $P(n)$ of a macrostate with n classical, and so distinguishable, particles in the rightmost chamber and $N - n$ particles in the leftmost chamber is

$$P(n) = \frac{N!}{n!(N-n)!}\left(\frac{1}{2}\right)^{N}$$

where the binomial coefficient $N!/\left[n!(N-n)!\right]$ is the multiplicity Ω of this macrostate and $(1/2)^{N}$ is the probability of any one microstate of this macrostate. A few empirical tests for small values of N suggest that the value of n that maximizes $P(n)$ is approximately $N/2$ – in which case the two chambers have equal numbers of particles.

A more systematic derivation of this result is possible when the numbers of particles in the two chambers, n and $N-n$, are so large as to be effectively continuous variables, that is, when $n \gg 1$ and $N-n \gg 1$. In this case Stirling's approximation $n! \approx n^{n}e^{-n}\sqrt{2\pi n}$ or, more crudely, $\ln n! \approx n \ln n - n$, obtains and allows us to use ordinary calculus to maximize, not $P(n)$, but rather $\ln P(n)$. Then

$$\ln P(n) = N \ln N - N - n \ln n + n - (N-n)\ln(N-n) + (N-n) - N \ln 2$$
$$= N \ln N - n \ln n - (N-n)\ln(N-n) - N \ln 2.$$

Setting the derivative of $\ln P(n)$ with respect to n equal to zero produces

$$-\ln n + \ln(N - n) = 0,$$

whose solution is $n = N/2$.

2.4 Statistical entropy and multiplicity

It may be surprising that a statistical version of the entropy is not given to us ready-made but, rather, must be constructed out of desirable properties and the laws of thermodynamics. To begin this construction, consider the irreversible processes illustrated in Figure 2.3. In the first sequence the particles of a gas are confined to one half of a space by an impenetrable barrier, the barrier is removed, and the gas particles fill the space available. In the second, two identical bodies with different temperatures, T_H and $T_C < T_H$, are separated by an adiabatic barrier, the barrier is removed, and the two bodies approach thermal equilibrium.

Dependence of entropy on macrostate multiplicity

In each of these irreversible transitions the entropy of a macrostate of an isolated system increases from an initial value S_i to a final value $S_f > S_i$ in accordance with the second law of thermodynamics. At the same time the number of microstates accessible to the isolated system increases from an initial value Ω_i to a final value $\Omega_f > \Omega_i$. This correlation between the evolution of the entropy S of an isolated system and the evolution of its multiplicity Ω suggests that the statistical entropy of a macrostate is an ever increasing, monotonic function $S(\Omega)$ of macrostate multiplicity. Assuming that such a function exists is the first step in constructing a definition of statistical entropy.

Boltzmann, in his landmark 1877 paper, suggested that the function $S(\Omega)$ is adequately expressed by the proportionality $S \propto \ln \Omega$. Later, Max Planck (1858–1947), starting from reasonable hypotheses, derived the relation, $S = k \ln \Omega$, where k is a universal constant. Here we follow and at the same time generalize Planck's construction. While our derivation is lengthy, we emerge with a result only slightly more complex than Planck's – that the entropy of an isolated system in a given macrostate is $S(\Omega) = c + k \ln \Omega$ where Ω is the macrostate multiplicity, k is a universal constant, and c is a constant that ensures that the entropy is an additive state variable.

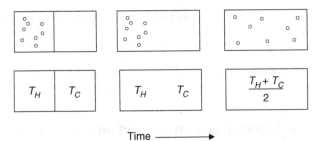

Time ———————➤

Figure 2.3 Top sequence: an impenetrable barrier is removed and the gas fills its container. Bottom sequence: an adiabatic barrier is removed and the two parts approach a common temperature.

Additivity

Additivity is the second property we incorporate into our definition of statistical entropy. Accordingly, the entropy S of a two-part composite system, illustrated in Figure 2.4, in which each part is itself an isolated system, denoted A and B, is *additive* in the sense that the entropy of the composite system is the sum of the entropy of its parts. Thus,

$$S_{A+B} = S_A + S_B, \qquad (2.1)$$

where $S_A{+}_B$ is the entropy of the composite system, S_A is the entropy of sub-system A, and S_B is the entropy of sub-system B. Adopting additivity (2.1) ensures that the statistical entropy so constructed will be consistent with the requirement, imposed in Section 1.5, on the additivity of the increments of thermodynamic entropy. The notation of (2.1) allows for sub-systems that, in general, may be quite different from one another as in, for example, a system composed of a bottle of helium gas in one corner of a laboratory and a crystal of salt in another.

Given that the statistical entropy is a function $S(\Omega)$ of macrostate multiplicity Ω, additivity (2.1) implies that

$$S_{A+B}\left(\Omega_{A+B}\right) = S_A\left(\Omega_A\right) + S_B\left(\Omega_B\right), \qquad (2.2)$$

where Ω_A is the multiplicity of a macrostate of sub-system A, Ω_B is the multiplicity of a macrostate of sub-system B, and Ω_{A+B} is the multiplicity of a macrostate of the composite system A+B. When the two subsystems are simply smaller copies of the composite system, the functions $S_A\left(\Omega_A\right)$ and $S_B\left(\Omega_B\right)$ and the multiplicities Ω_A and Ω_B must be similarly structured and *additivity* reduces to *extensivity*.

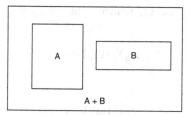

Figure 2.4 Isolated composite system A+B consisting of two isolated subsystems, A and B.

Independence

Thirdly, we assume that isolated sub-systems, A and B, independently realize their macrostates. Therefore, the multiplicity of the composite system is given by

$$\Omega_{A+B} = \Omega_A \Omega_B \tag{2.3}$$

and condition (2.2) becomes

$$S_{A+B}\left(\Omega_A\Omega_B\right) = S_A\left(\Omega_A\right) + S_B\left(\Omega_B\right). \tag{2.4}$$

Actually, the very isolation of the two subsystems implies the independence of their macrostates. This result (2.4) incorporates all three of the following conditions: the dependence of entropy on macrostate multiplicity $S(\Omega)$, additivity (2.1), and the independence of isolated subsystems (2.3).

Exploiting derivatives

We now take derivatives of expression (2.4) with respect to the macrostate multiplicities Ω_A and Ω_B – a tactic that reduces a stronger condition (additivity of entropies) to a weaker one (additivity of entropy increments). If we want to restore the full content of additivity, we must reimpose (2.4) on the result of our derivation. The two partial derivatives of (2.4) yield

$$S'_{A+B}\left(\Omega_A\Omega_B\right)\Omega_B = S'_A\left(\Omega_A\right) \tag{2.5a}$$

and

$$S'_{A+B}\left(\Omega_A\Omega_B\right)\Omega_A = S'_B\left(\Omega_B\right), \tag{2.5b}$$

where a prime symbol indicates a derivative with respect to the indicated argument. For example, $f'(x) = df/dx$. Eliminating $S'_{A+B}\left(\Omega_A\Omega_B\right)$ from (2.5a) and (2.5b) produces

$$\Omega_A S'_A\left(\Omega_A\right) = \Omega_B S'_B\left(\Omega_B\right). \tag{2.6}$$

Since the variables Ω_A and Ω_B realize their values independently, Eq. (2.6) implies that

$$\Omega_A S'_A(\Omega_A) = k \tag{2.7a}$$

and

$$\Omega_B S'_B(\Omega_B) = k, \tag{2.7b}$$

where, formally, k is a separation constant. Integrating (2.7a) and (2.7b) produces

$$S_A(\Omega_A) = c_A + k \ln \Omega_A \tag{2.8a}$$

and

$$S_B(\Omega_B) = c_B + k \ln \Omega_B, \tag{2.8b}$$

where c_A and c_B are integration constants.

The meaning of "constant"

We pause to consider the nature of the three constants, k, c_A, and c_B, generated by solving the coupled differential equations (2.5a) and (2.5b). The word *constant* should always prompt us to ask: "Constant with respect to what?" In this case, the answer is: "Constant with respect to the variables in terms of which the derivatives in Eqs. (2.5)–(2.7) are defined, that is, constant with respect to the macrostate multiplicities Ω_A and Ω_B."

The constant k is additionally constrained. Since k is associated with both sub-systems A and B (actually, any number of subsystems), k is independent of the parameters that identify any particular macrostate of any particular system or sub-system. In brief, k is a universal constant. Max Planck was the first to emphasize the universality of k. In his honor H. A. Lorentz (1853–1928), for a time, called k *Planck's constant*. However, the phrase *Planck's constant* is now reserved for another of Planck's discoveries, and we refer to k as *Boltzmann's constant*. We shall find that $k = R/N_A$, where R is the misleadingly named "gas constant" and N_A is Avogadro's number. In SI units the value of Boltzmann's constant is given by

$$k = 1.38 \times 10^{-23} \frac{\text{kg} \cdot \text{m}^2}{\text{K} \cdot \text{s}^2}. \tag{2.9}$$

Thus, the units of Boltzmann's constant k are those of energy over thermodynamic temperature, that is, k has the units of entropy.

The statistical entropy of an isolated system

If Eqs (2.8a) and (2.8b) are true of any isolated system, then

$$S_{A+B}\left(\Omega_{A+B}\right) = C_{A+B} + k \ln \Omega_{A+B} \qquad (2.10)$$

applies to a composite system composed of sub-systems A and B. Additivity (2.2) and independence (2.3), or equivalently (2.4), then require that the three constants, c_{A+B}, c_A, and c_B, are related by

$$c_{A+B} = c_A + c_B. \qquad (2.11)$$

Therefore, when we apply the general result

$$S(\Omega) = c + k \ln \Omega \qquad (2.12)$$

to a composite system and to each of its two isolated parts, we should always be ready to adopt (2.11) in order to restore the full content of additivity to our definition of statistical entropy.

Alternate forms

We can express the general result (2.12) in a revealing form by writing it twice, once for each of two different macrostates of the same system, an initial state denoted i and a final state denoted f, and subtracting the two equations from each other. In this way we find that

$$S(\Omega_i) = c + k \ln \Omega_i \qquad (2.13)$$

and

$$S(\Omega_f) = c + k \ln \Omega_f \qquad (2.14)$$

imply

$$S(\Omega_f) - S(\Omega_i) = k \ln \left(\frac{\Omega_f}{\Omega_i} \right). \qquad (2.15)$$

This result (2.15) is especially useful in describing the entropy increment of an isolated system undergoing an irreversible change.

In particular, Eq. (2.15) allows us to restate the second law of thermodynamics in the language of multiplicity: *Because an isolated system may relax but may never impose an internal constraint, the multiplicity of the final macrostate of an isolated system is at least as large as the multiplicity of its initial macrostate.* Some texts characterize the irreversible transition from an internally constrained system to one less constrained as one from a relatively ordered macrostate to one that is less ordered. Actually, Boltzmann never used the words *order* and *disorder* (German: *Ordnung* and *Unordnung*) in his 1877 paper. And, when applied to the relative entropies of a system, the opposites *order* and *disorder* sometimes mislead.

Alternatively, we can replace the initial macrostate of (2.15) Ω_i with a special reference macrostate Ω_o and the final macrostate multiplicity Ω_f with an arbitrary macrostate multiplicity Ω. Then (2.15) assumes a form,

$$S(\Omega) = S(\Omega_o) + k \ln\left(\frac{\Omega}{\Omega_o}\right), \qquad (2.16)$$

useful for describing the entropy $S(\Omega)$ of an arbitrary macrostate relative to the entropy $S(\Omega_o)$ of a special reference macrostate.

The entropy of a non-isolated system

By carefully deriving (2.12), (2.15), and (2.16) we discover exactly what ideas are contained in these expressions for statistical entropy and what ideas are not. In particular: the statistical entropy of an isolated system is a function of macrostate multiplicity, is an additive function of state, and reflects the independence of isolated parts.

Recall, though, a basic thermodynamic truth: everything known about the thermodynamic behavior of a system including all its equations of state is encapsulated in how its entropy depends upon its extensive variables. This is so even when the special assumptions used (isolating the system and positing the existence of independent subsystems) in order to determine how the statistical entropy depends upon macrostate multiplicity are not observed.

For instance, in the applications we explore throughout much of this book we consider a closed, N-particle, simple fluid system with extensive variables: internal energy E, volume V, and particle number N. However, since the number of particles N in a closed system is conserved, its macrostates are a function only of E and V. In this case, $\Omega = \Omega(E, V, N)$, $c = c(N)$, and the general result (2.12) can be expressed as

$$S(E, V, N) = c(N) + k \ln \Omega(E, V, N). \qquad (2.17)$$

Thus, once we determine the multiplicity function $\Omega(E, V, N)$ and, by requiring additivity, the "constant" $c(N)$, we know the entropy function $S(E, V, N)$. And once we know the entropy function we know, through its derivatives, the system's equations of state. Furthermore, these equations of state apply in all situations – not only when the system is closed and isolated. They describe, for instance, how the system changes when heated or cooled, when performing work or being worked upon. To summarize: The presumption of an isolated system allows us to determine the entropy function and once we know this function we need no longer maintain this presumption.

The limits of classical statistical thermodynamics

The arguments of this section do not allow us to determine the absolute entropy of any macrostate. An expression for absolute entropy is one that, given complete information about a particular macrostate, provides a determinate value of the entropy of that macrostate. Equations (2.15) and (2.16) obviously produce only relative entropies. And only relative entropies emerge from the first and second laws of thermodynamics.

This limitation to relative entropies originates with our inability, given classical presuppositions, to *uniquely* identify and count microstates. Furthermore, reference microstates in classical statistical mechanics are necessarily arbitrary and so classical macrostates are also necessarily arbitrary. Only the introduction of quantum physics allows us to uniquely identify microstates. But, of course, Boltzmann and the other classical statistical mechanicians were unaware of the possibilities inherent in quantum physics.

What Boltzmann did know was how to arbitrarily discretize space in such a way as to construct the arbitrary microstates we call *classical microstates*. Boltzmann then applied the fundamental postulate to these arbitrarily constructed classical microstates by asserting that $S \propto \ln \Omega$. Boltzmann's method of discretizing phase space allowed him and will allow us to determine classical versions of statistical entropies from which it is possible to derive equations of state. Of course we will use $S(\Omega) = c + k \ln \Omega$, in place of Boltzmann's $S \propto \ln \Omega$ in order to ensure that the entropy is always an additive function of state variables.

Example 2.2 Joule expansion

Problem: Consider, as in Example 2.1, a system with left and right, equal-volume chambers containing a total of N identical but classical, and so distinguishable, particles. Suppose this system is initialized in a macrostate in which all its particles are contained within its left chamber. Then the particles are allowed to move freely between both chambers and achieve a final equilibrium macrostate in which the particles are distributed throughout both chambers as illustrated in Figure 2.5. How much does the entropy of the system increase?

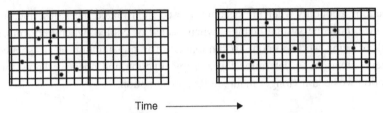

Time ─────────────▶

Figure 2.5 Joule expansion. After the barrier that keeps the particles in the left hand side of two equal-volume chambers is removed, the particles occupy both chambers.

Solution: The number of spatial microstates available to a single particle increases by a factor of 2 during this process. And the number of spatial microstates available to two distinguishable particles increases by 2×2. Therefore, the multiplicity of the N-particle system increases by 2^N during this Joule expansion. Consequently, the ratio of the final to the initial macrostate multiplicity is

$$\frac{\Omega_f}{\Omega_i} = 2^N,$$

and the entropy is incremented by

$$S_f - S_i = k \ln\left(\frac{\Omega_f}{\Omega_i}\right),$$

$$\Delta S = k \ln\left(\frac{\Omega_f}{\Omega_i}\right)$$

$$= Nk \ln 2.$$

Apparently each particle contributes $k \ln 2$ to the entropy increment.

Example 2.3 A paradox

Problem: Consider a system of $2N$ identical, classical particles initialized with N particles in the left of two equal-volume chambers and N in the right. The particles are allowed to move freely and achieve a final, equilibrium state. What is the entropy increment?

Solution: One response is to say that the entropy of the N particles initially in the left chamber increases, as in Example 2.2, by $Nk \ln 2$ as those particles spread throughout both chambers and likewise for the particles initially in the right chamber. In this way the total entropy would increase by $2Nk \ln 2$.

But this is absurd. For, if we were then to reinsert a barrier between the two chambers, remove that barrier, and allow the particles on each side to again mix with each other, the system entropy would again increase by $2Nk\ln 2$ and so on.

The key to resolving this paradox is to recall that a system's entropy only increases when the multiplicity of its final macrostate is larger than the multiplicity of its initial macrostate. In this problem the initial and final macrostates are the same because they have the same description: N classically identical molecules occupy each chamber. Therefore the initial and final macrostate multiplicities and entropies are the same, and there is no entropy increment. True, the final microstate of the system may differ from its initial microstate, but this is of no consequence for entropy. The entropy increment is a function only of macrostate multiplicity and not of the particular microstates that compose the macrostate.

Example 2.4 Entropy of mixing

Problem: A system consisting of $2N$ classical particles is initialized in a macrostate in which N particles of one kind, for example, nitrogen molecules, are all in the left of two equal-volume chambers and N particles of another kind, for example, oxygen molecules, are all in the right chamber. The particles are allowed to mix. What is the entropy increment?

Solution: The microstates available to each kind of particle are doubled during this process and this doubling does increase the number of microstates available to the system in its final macrostate. Therefore, the initial and final multiplicities of the $2N$-particle system are in the ratio

$$\frac{\Omega_f}{\Omega_i} = 2^{2N}.$$

Accordingly the entropy increment $\Delta S = k\ln\left(\Omega_f/\Omega_i\right)$ of the system is

$$\Delta S = 2Nk\ln 2.$$

2.5 Maxwell's demon

While Boltzmann was the first to suggest that a relation exists between the entropy S of an isolated system and its macrostate multiplicity Ω, it was James Clerk Maxwell (1831–1879) who first suggested that entropy is a probabilistic concept – first privately in a letter to his friend Peter Guthrie Tait in 1867 and

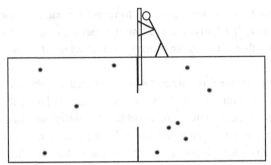

Figure 2.6 Maxwell's demon. By opportunely opening and closing a door connecting two chambers of gas, the demon can increase the temperature of the gas in
one chamber over that of the other.

later in his 1871 text *Theory of Heat*. Maxwell died before he could develop
this idea, but his way of arguing for it had important consequences for the
way in which the concept of statistical entropy has developed. (See the entry
Maxwell's Demon in the Annotated Further Reading.)

Maxwell's conceit was to imagine a small, "neat fingered" being, later called
a demon and much later called *Maxwell's demon*, who attended a door between
two chambers of air. (See Figure 2.6.) The air in the two chambers initially has
the same temperature and density, but even so its individual molecules move
at various speeds – a phenomenon first discovered by Maxwell. The demon's
job is to open the door in order, say, to allow the fastest moving particles in
the right chamber to move to the left chamber and to allow the slowest moving particles in the left chamber to move to the right chamber. In this way, the
demon could cause the air in the left chamber to become hotter than the air in
the right chamber while keeping their densities the same. If we consider the
two chambers, the air they contain, and the demon as an isolated system, this
process directly violates Clausius's version of the second law assuming the
entropy of the demon does not change. Admittedly, the latter is a significant
assumption.

Maxwell claimed that his demon-inhabited thought experiment demonstrated that the second law of thermodynamics "has only statistical certainty,"
that the second law can at times be violated, and that the statistical entropy
of an isolated system can momentarily decrease. For if a demon can cause
the entropy of an isolated system to decrease, possibly an inanimate automaton like a spring-loaded, one-way door can do the same. Indeed, subsequent
analyses and numerical simulations of many cleverly constructed, inanimate
"demons" have shown that the statistical entropy of an isolated system may
fluctuate – but only fluctuate. And the larger the system, the smaller is the relative size of these fluctuations.

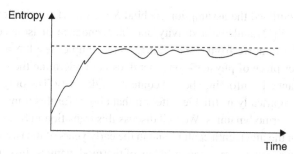

Figure 2.7 When an internal constraint of an isolated system is relaxed, the system entropy (solid line) approaches and fluctuates close to and just under its maximally probable, equilibrium value (dotted line).

The entropy of a non-equilibrium system

The machinations of Maxwell's demon suggest a broader interpretation of the relative entropy according to which the multiplicity Ω is that of an instantaneous macrostate as a system evolves from an initial low entropy reference macrostate toward an equilibrium macrostate. This interpretation assumes that the system always occupies a macrostate at every instant of its evolution. In this case, $S(\Omega)$ where

$$S(\Omega) = S(\Omega_o) + k \ln\left(\frac{\Omega}{\Omega_o}\right) \tag{2.18}$$

is the instantaneous entropy of a system, not necessarily in equilibrium, relative to the entropy $S(\Omega_o)$ of the system in a reference macrostate. Generally, the evolution of an isolated system is in the direction of increasing entropy. But around this general trend Maxwell's demon teaches us there may be small fluctuations. As illustrated in Figure 2.7, the instantaneous macrostate multiplicity Ω and hence the instantaneous entropy $S(\Omega)$ fluctuate as the system occupies different, not quite optimal, macrostates.

2.6 Relative versus absolute entropy

The entropy relation $S = c + k \ln \Omega$ is perfectly adequate and, in fact, necessary for describing the entropy of a classical system. We will explore two such systems in the next chapter: the ideal gas in Sections 3.1–3.5 and the ideal solid in Section 3.6. Einstein and others recognized the general validity of $S = c + k \ln \Omega$. Nevertheless, one usually finds Max Planck's more succinct result

$$S = k \ln \Omega \tag{2.19}$$

in textbooks.

We have outlined the assumptions behind $S = c + k \ln \Omega$ in Section 2.4: the functionality $S(\Omega)$ and the additivity and independence of isolated subsystems. In order to move from $S = c + k \ln \Omega$ to the simpler result $S = k \ln \Omega$ we need a further piece of physics – one that does not undermine the role played by the constant c in enforcing the full content of additivity. The only generally applicable, empirically justifiable criterion that allows for this transition is the third law of thermodynamics. We will discuss this transition in Section 4.5.

But late in the nineteenth century and in the early years of the twentieth century Planck was not aware of the third law of thermodynamics. Instead Planck, in addition to assuming the existence of a function $S(\Omega)$, additivity (2.1), and independence (2.3), implicitly, and without a priori justification, assumed that $S(\Omega)$ is a universal function where here "universal" means a function only of the multiplicity Ω and universal constants. The universality of $S(\Omega)$, additivity, and independence produce the result $S = k \ln \Omega + c$ where both k and c are universal constants. If c is a universal constant then c is the same when applied to a system and its two isolated parts. In this case $c_{A+B} = c_A + c_B$ reduces to an equation $c = c + c$ whose only solution is $c = 0$. Thus Planck reduced $S = k \ln \Omega + c$ to $S = k \ln \Omega$. But in the process Planck eliminated his ability to impose the full content of a property he had originally assumed: additivity.

Planck's method actually worked for him because it anticipated future developments – in particular, the formulation and empirical justification of the third law of thermodynamics and the advent of quantum physics. Together these developments ensured that $S(\Omega)$ can be both a universal function of multiplicity Ω and an additive state variable. But some classical systems, including the ideal gas system, are neither third law compliant nor quantum mechanically conceived and for this reason are not well served by Planck's equation $S = k \ln \Omega$. For these classical systems we will use the relative entropy $S = c + k \ln \Omega$ in place of the absolute entropy $S = k \ln \Omega$.

Problems

2.1 Probabilities

A container is divided into two compartments with volumes in the ratio $V_1/V_2 = 2$. Volume V_1 contains 1000 N_2 molecules and volume V_2 contains 100 O_2 molecules. (a) The interior wall separating the two volumes is punctured and the gases are allowed to achieve an equilibrium macrostate. What is the increase in the entropy of the two-gas system? (b) What is the probability that the gases will return to their original configuration?

2.2 Playing card multiplicities

(a) How many ways are there of drawing a 3 from a complete 52-card deck of playing cards? (b) How many ways are there of drawing a diamond from a complete 52-card deck? (c) How many ways are there of drawing a 3 from a complete deck of cards followed by a diamond drawn from a complete deck of cards? (d) What can you conclude about the dependence or the independence of the two events: "Draw a 3 from a complete deck of cards" and "Draw a diamond from a complete deck of cards?"

2.3 Stirling's approximation

What is the percent difference between $\ln 10!$ and Stirling's approximate value $10 \ln 10 - 10$? How large must n be in order that the percent difference between $\ln n!$ and $n \ln n - n$ be less than 1%?

2.4 The art of counting

Because the multiplicity Ω is the number of microstates available to a system in a given macrostate, we will, in the chapters that follow, often be presented with the problem of counting microstates. The following exercises introduce us to three different ways of counting microstates. You might try generating these formulas first by doing the counting when N and n are small numbers. (a) How many ways can N distinguishable particles be placed in n different boxes or cells? Find this number when $N = 50$ and $n = 100$. (b) How many ways can N indistinguishable particles be placed in n different boxes or cells? Find this number when $N = 50$ and $n = 100$. (c) How many ways can N indistinguishable particles be placed in $n [\geq N]$ different boxes or cells when no more than one particle can occupy a box? Find this number when $N = 50$ and $n = 100$. (d) Which is larger: the answer to question (b) or the answer to question (c)? Explain why your answer makes sense.

2.5 Rubber elasticity

Consider the following simple, one-dimensional, statistical model of a rubber band. Suppose a length L of a band of rubber consists of $N \; [\gg 1]$ links where each link has length a. Some n_+ of these links extend in the positive x direction while $n_- \; [= N - n_+]$ extend in the negative x direction. The orientation of the links contributes to the total length $L = a(n_+ - n_-)$ of the rubber band but does not change its energy. Therefore, a macrostate is defined by a length L. (a) First, determine the multiplicity Ω in terms of the length L. Use "N choose n_+." (b) Then determine the entropy in terms of L. (c) Justify the statement that the force F with which a length L of rubber pulls is related to the entropy S by $F/T = (\partial S/\partial L)$ by inspecting the system's fundamental constraint $0 = T\,dS - F\,dL$. Use this relation to determine an expression for the force F in terms of T and L. (d) Show that a linear, Hooke's law relationship between F and L follows whenever $L/Na \ll 1$.

3

Entropy of classical systems

3.1 Ideal gas: volume dependence

Boltzmann worked on the statistical mechanics of ideal gases for most of his professional life. We devote the next five sections to this task. Our ultimate goal is to construct an ideal gas entropy $S(E,V,N)$ that is an extensive and so additive function of its extensive variables: internal energy E, volume V, and particle number N, and whose partial derivatives with respect to E and V produce the ideal gas equations of state: $E = C_v T$ and $PV = NkT$. Here we address the volume dependence and in the next section the conjoined volume and energy dependence of this entropy.

We first employ a version of the relative entropy

$$S(E,V,N) = c(E,N) + k \ln \Omega(V) \qquad (3.1)$$

that applies to a fluid system whose macrostate multiplicity is a function only of the system volume V. The positional microstates that compose the macrostates of this system are, in general, described by identifying the $3N$ position variables: $x_1, y_1, z_1, x_2, y_2, z_2, \ldots, x_N, y_N$, and z_N, that locate the N particles of the system.

In order to construct classical microstates that will help us determine the multiplicity Ω we divide this $3N$-dimensional space into small regions or cells of size $\delta x_1 \delta y_1 \delta z_1 \delta x_2 \cdots \delta z_N$ that are uniform and symmetric, that is, for which $\delta x_1 = \delta y_1 = \delta z_1 = \delta x_2 = \cdots = \delta z_N = \delta s$. Thus, each cell in this $3N$-dimensional position space has volume $\delta s^{3N} = \delta V^N$ and each identifies a different classical microstate. There are $(V/\delta V)^N$ such cells. A single point representing the N-particle system moves through this space and among these cells.

Because there are $(V/\delta V)^N$ different cells accessible to the N-particle system contained within volume V, the system multiplicity is given by $\Omega = (V/\delta V)^N$. Consequently, (3.1) becomes

$$S(E,V,N) = c(E,N) + Nk \ln\left(\frac{V}{\delta V}\right). \tag{3.2}$$

The fundamental relation of a simple fluid, $dS = dE/T + PdV/T$, implies that $P/T = (\partial S/\partial V)_{E,N}$ and so, given (3.2), that

$$PV = NkT, \tag{3.3}$$

which, of course, is the ideal gas pressure equation of state. Note that we have calculated the entropy function of an ideal gas only to the extent that from its volume derivative we can determine its pressure equation of state. Because we have not yet determined the complete dependence of the entropy on all its variables, we cannot yet derive all of the system's equations of state.

3.2 Ideal gas: volume and energy dependence

In deriving an expression for the entropy $S(E,V,N)$ that depends on both the internal energy E and the volume V of an N-particle ideal gas we generalize the pattern of argument established in the previous section. The structure of the classical microstates available to this system is now more complex because it consists of both a particle position space and a particle velocity v or momentum $p\ [\equiv mv]$ space. Boltzmann realized that, because position and momentum coordinates play parallel roles in the Hamiltonian formulation of classical mechanics, microstates are best described in terms of position and momentum variables.

Discretizing phase space

The N-particle, monatomic ideal gas system is described in terms of $3N$ position $x_1, y_1, z_1, \ldots, z_N$ and 3N momentum $p_{x,1}, p_{y,1}, p_{z,1}, \ldots, p_{z,N}$ coordinates. Together these coordinates define a $6N$-dimensional *phase space* through which a single point representing the N-particle system moves. We discretize this $6N$-dimensional phase space in order to define the classical microstates and determine the multiplicity of the macrostate. Accordingly, we divide the position space into uniform, symmetric cells of size $\delta x_1 \delta y_1 \delta z_1 \delta x_2 \cdots \delta z_N = \delta s^{3N}$ and the momentum space into uniform, symmetric cells of size $\delta p_{x,1} \delta p_{y,1} \delta p_{z,1} \delta p_{x,2} \cdots \delta p_{z,N} = \delta p^{3N}$. We also adopt the following notation,

$$\delta x_1 \delta y_1 \delta z_1 \delta x_2 \cdots \delta z_N \delta p_{x,1} \delta p_{y,1} \delta p_{z,1} \delta p_{x,2} \cdots \delta p_{z,N} = (\delta s \delta p)^{3N}$$
$$= H^{3N}, \tag{3.4}$$

where $H \left[= \delta s \, \delta p \right]$ is a quantity small enough to make the multiplicity of the system large compared to 1 but is otherwise arbitrary. The dimensions of H are those of action and also of angular momentum, that is, of energy multiplied by time with SI units of $\mathrm{kg \cdot m^2/s}$. In Chapter 4 we will find that, according to the requirements of quantum mechanics, the arbitrary value of H is fixed by a universal, determinate constant h – one we now call Planck's constant.

Identifying the cell in $6N$-dimensional phase space within which lie the $6N$ coordinates of an ideal gas roughly identifies a classical microstate. We say "roughly" because one phase space cell necessarily encompasses many different points. The degree of roughness is determined by the size of H. The larger H, the rougher the description.

The multiplicity

Because the position and momentum coordinates of an ideal gas system are constrained in different ways, they do not play entirely parallel roles. For instance, the position coordinates must be consistent with locating the system particles within a volume V. For this reason the number of allowed position space cells, that is, the multiplicity associated with position space, is

$$\left(\frac{V}{\delta s^3} \right)^N . \tag{3.5}$$

In contrast, the squares of the momentum coordinates must sum to the constant $2mE$ so that

$$p_{x,1}^2 + p_{x,2}^2 + p_{z,1}^2 + p_{x,2}^2 + p_{y,2}^2 + p_{z,2}^2 + \cdots + p_{x,N}^2 + p_{y,N}^2 + p_{z,N}^2 = 2mE. \tag{3.6}$$

where E is the system energy. Just as $x^2 + y^2 + z^2 = R^2$ defines a 3-dimensional sphere of radius R, the constraint (3.6) defines a $3N$-dimensional sphere in momentum space of radius $\sqrt{2mE}$. The number of momentum space cells through which the system may move is the volume of the cells that lie on the surface of this sphere divided by the volume δp^{3N} of one such cell. And the volume of the cells that lie on the surface of the $3N$-dimensional sphere equals its surface area times a momentum space cell width δp.

This surface area of a $3N$-dimensional sphere can be calculated in the following way. Just as the surface area of a 3-dimensional sphere $4\pi R^2$ is the derivative d/dR of the volume of a 3-dimensional sphere $4\pi R^3/3$, the surface area of a $3N$-dimensional sphere $A_{3N}(R)$ is the derivative d/dR of the volume of a $3N$-dimensional sphere $V_{3N}(R)$, that is,

$$A_{3N}(R) = \frac{d}{dR} V_{3n}(R). \tag{3.7}$$

Furthermore, the volume $V_{3N}(R)$ of a $3N$-dimensional sphere of radius R must be proportional to R^{3N}, that is,

$$V_{3N}(R) = C_{3N} R^{3N}, \tag{3.8}$$

where C_{3N} is a number independent of V and E. For example $C_3 = 4\pi/3$. In general,

$$C_n = \frac{\pi^{n/2}}{[n/2+1]!}. \tag{3.9}$$

Since we not concerned with cases for which n is a small integer, we replace $(n/2+1)!$ in (3.9) with $(n/2)!$. And we apply Stirling's approximation,

$$\ln n! \approx n \ln n - n \tag{3.10}$$
$$\approx \ln\left(\frac{n}{e}\right)^n$$

or its equivalent

$$\left(\frac{n}{2}\right)! \approx \left(\frac{n}{2e}\right)^{n/2} \tag{3.11}$$

to expression (3.9). Therefore,

$$C_n \approx \left(\frac{2e\pi}{n}\right)^{n/2} \tag{3.12}$$

when $n \gg 1$. Using (3.12) in (3.8) and (3.7) we find that

$$A_{3N}(R) = \left(\frac{2e\pi}{3N}\right)^{3N/2} \frac{d}{dR}\left(R^{3N}\right)$$
$$= \left(\frac{2e\pi}{3N}\right)^{3N/2} (3N) R^{3N-1}. \tag{3.13}$$

Therefore, the volume of the momentum space through which the system may move is the volume $(2e\pi/3N)^{3N/2}(3N) R^{3N-1} \delta p$ of a one-cell wide shell of radius $R = \sqrt{2mE}$. Consequently, the number of allowed momentum-space cells in this shell is

$$(2e\pi)^{3N/2}(3N)^{1/2}\left(\frac{2mE}{3N}\right)^{(3N-1)/2}\left(\frac{\delta p}{\delta p^{3N}}\right) \tag{3.14}$$

where we have replaced R with $\sqrt{2mE}$.

The multiplicity $\Omega(E,V,N)$ of the N-particle, monatomic, ideal gas with energy E and volume V is the product of the number of cells available in position space (3.5) and the number of cells available in momentum space (3.14). Therefore,

$$
\begin{aligned}
\Omega(E,V,N) &= \left(\frac{V}{\delta V}\right)^{N} (2e\pi)^{3N/2} (3N)^{1/2} \left(\frac{2mE}{3N}\right)^{(3N-1)/2} \left(\frac{\delta p}{\delta p^{3N}}\right) \\
&= \left(\frac{V}{\delta s^3}\right)^{N} (2e\pi)^{3N/2} \left(\frac{2mE}{3N\,\delta p^2}\right)^{3N/2} \\
&= \left[V\left(\frac{E}{N}\right)^{3/2} \left(\frac{4e\pi m}{3H^2}\right)^{3/2}\right]^{N},
\end{aligned}
\tag{3.15}
$$

where in the first step of this equation sequence we have, since $N \gg 1$, dropped terms of order 1 compared to terms of order N. The terms dropped do not change the dimensionality of Ω. We have, in the second step, also adopted the notation $H = \delta s\, \delta p$.

The entropy of an ideal gas

Accordingly, the entropy of an N-particle, monatomic, ideal gas with energy E and volume V is

$$
S(E,V,N) = c(N) + k\ln\left[V\left(\frac{E}{N}\right)^{3/2} \left(\frac{4\pi em}{3H^2}\right)^{3/2}\right]^{N}.
\tag{3.16}
$$

The two equations of state follow from the derivatives, $(\partial S/\partial E)_{V,N} [= 1/T]$ and $(\partial S/\partial V)_{E,N} [= P/T]$, implied by the fundamental relation $dS = dE/T + P\,dV/T$. Given the entropy (3.16) we find that, as expected,

$$
E = \frac{3}{2}NkT
\tag{3.17}
$$

and

$$
PV = NkT.
\tag{3.18}
$$

(Of course, the energy of a system is known only within a constant. But if we replace E on the right hand side of (3.16) with $E + E_o$ and use $1/T = (\partial S/\partial E)_{V,N}$ and $P/T = (\partial S/\partial V)_{E,N}$, we again generate the familiar ideal gas equations of state (3.17) and (3.18).)

Summary

We summarize the method employed in these last two sections as follows. We start with the relative entropy $S[=c+k\ln\Omega]$ of an N-particle ideal gas whose macrostate multiplicity Ω depends upon the system internal energy E, volume V, and particle number N. Then we discretize the available $6N$-dimensional phase space and count the number of classical microstates through which the single point representing the system can move, that is, we determine the multiplicity Ω. This procedure produces the entropy $S(E,V,N)$ relative to that of a reference macrostate.

Both the multiplicity Ω and the entropy S determined in this way depend upon the size of the cells into which the phase space is divided. This cell size, in turn, depends upon an arbitrarily constructed quantity $H[=\delta s\,\delta p]$. Fortunately, none of the quantities that contain H can be measured, and none of the quantities that can be measured, for instance P, V, and T, depend on H. In classical statistical mechanics H is a mere methodological artifact with no empirical consequences.

3.3 Imposing extensivity

But more is required of the ideal gas entropy function (3.16). Any part of an ideal gas should itself be an ideal gas. In a word, the ideal gas should be *extensive*. For instance, if we divide the volume, the energy, and the number of ideal gas particles in half, we should create another ideal gas with half the entropy. In order that the entropy function (3.16) be extensive $S(E,V,N)$ must be a linear, homogeneous function of its extensive variables, that is,

$$S(\lambda E,\lambda V,\lambda N)=\lambda S(E,V,N) \tag{3.19}$$

for arbitrary λ.

Recall that the purpose of the "constant" $c(N)$ is to impose additivity, in this case extensivity, on the entropy. Together (3.16) and (3.19) require that

$$S(\lambda E,\lambda V,\lambda N)=\lambda S(E,V,N),$$

$$c(\lambda N)+k\lambda N\left\{\frac{3}{2}+\ln\left[\lambda V\left(\frac{E}{N}\right)^{3/2}\left(\frac{4\pi m}{3H^2}\right)\right]\right\}=\lambda c(N)$$

$$+k\lambda N\left\{\frac{3}{2}+\ln\left[V\left(\frac{E}{N}\right)^{3/2}\left(\frac{4\pi m}{3H^2}\right)^{3/2}\right]\right\}, \tag{3.20}$$

$$c(\lambda N)+k\lambda N\ln\lambda=\lambda c(N).$$

Equation (3.20) has a family of solutions,

$$c(N) = k(pN - N \ln N),$$ (3.21)

where p is any constant real number. The particular choice $p = 1$ turns the entropy (3.16) into

$$S(E,V) = kN \left\{ \frac{5}{2} + \ln \left[\left(\frac{V}{N} \right) \left(\frac{E}{N} \right) \right]^{3/2} \left(\frac{4m\pi}{3H^2} \right)^{3/2} \right\},$$ (3.22)

an historically important function that has all the properties one can reasonably expect of the classical, monatomic, ideal gas entropy. As before the parameter H that determines the size of a cell in phase space is a methodological artifact with no empirical consequences.

That the quantities p and H are arbitrary underscores what we already know: that the principles of a purely classical statistical mechanics based as they are on the first and second laws of thermodynamics do not invariably produce entropy functions that determine a unique value in terms of its extensive variables E, V, and N. We say "invariably" because imposing extensivity on blackbody radiation, as discussed below, does produce an absolute entropy with a unique value. But blackbody radiation is a particularly fundamental system that, for this reason, became an object of special interest in the late nineteenth century.

Example 3.1 Extensivity of blackbody radiation

Problem: Can the entropy of blackbody radiation be made extensive?

Solution: According to (1.38) the entropy of blackbody radiation is described by

$$S(E,V) = \frac{4a^{1/4}V^{1/4}E^{3/4}}{3} + b$$

where a is the radiation constant and b is an arbitrary constant independent of the system's thermodynamic variables E and V. Extensivity requires that

$$S(\lambda E, \lambda V) = \lambda S(E,V),$$

that is,

$$\frac{4a^{1/4}(\lambda V)^{1/4}(\lambda E)^{3/4}}{3} + b = \frac{\lambda 4a^{1/4}V^{1/4}E^{3/4}}{3} + \lambda b,$$

for arbitrary λ. A vanishing b is the only value consistent with an arbitrary value of λ. Therefore, $b = 0$ and

$$S(E,V) = \frac{4a^{1/4}V^{1/4}E^{3/4}}{3}$$

is the extensive entropy of blackbody radiation.

3.4 Occupation numbers

Although the ideas behind our derivation of the entropy of an ideal gas are simple ones, their application depends upon the unfamiliar geometry of $6N$-dimensional phase space. Boltzmann favored another derivation that employs a single 6-dimensional phase space common to each of the system's N particles. The reduction of $6N$- to 6-dimensional phase space is possible whenever the particles that compose the system occupy their positions in a common phase space independently of one another except for system-wide constraints such as those imposed on the total energy E and particle number N.

Figure 3.1 illustrates Boltzmann's idea for a positional phase space volume V common to all 10 particles of this model system. This positional phase space is divided into uniform cubes of size δV that define a single positional microstate of a single particle. Thus, there are $V/\delta V$ such single-particle positional microstates. The classical microstate of this 10-particle system is identified by the particular arrangement of the 10 distinguishable particles in the common set of cells.

More generally, Boltzmann employed special macrostates defined by the numbers of particles n_j that occupy the jth cell $[j = 1, 2, ..., J]$ of the 6-dimensional, single-particle, phase space. The set of *occupation numbers*, $n_1, n_2, ..., n_J$, defines a particular macrostate of the system. The occupation numbers that maximize the multiplicity Ω of this macrostate also maximize the entropy $S(\Omega)$. This most probable macrostate defines an equilibrium macrostate. The use of special macrostates described by occupation numbers might seem an unnecessary complication. But occupation numbers allow us to discover an important result that generally applies to many systems including both the ideal classical gas and the ideal classical solid that we explore in the following two sections. This important result is the *Maxwell–Boltzmann distribution*.

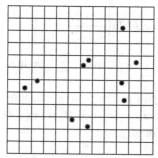

Figure 3.1 The volume V is divided into $V/\delta V$ cells each of size δV. One arrangement of 10 particles is shown.

The Cartesian coordinates of a single-particle, 6-dimensional, phase space common to all particles are: x, y, z, p_x, p_y, and p_z. We divide this phase space into J uniform, symmetric cells of size

$$\delta x\,\delta y\,\delta z\,\delta p_x\,\delta p_y\,\delta p_z = \left(\delta s\,\delta p\right)^3$$
$$= H^3 \tag{3.23}$$

where $\delta s = \delta x = \delta y = \delta z$, $\delta p = \delta p_x = \delta p_y = \delta p_z$, and $H = \delta s\,\delta p$, and label these cells with index $j = 1, 2, ..., J$.

The system occupies a macrostate in which of the N distinguishable particles n_1 are in the first cell, n_2 are in the second cell, ..., and n_J are in the Jth phase space cell. Imagine placing these N distinguishable particles in cells, one after another. We could do this by arranging these particles in a row, the first n_1 of which go in the first cell, the next n_2 of which go in the next cell and so on. Since there are $N!$ ways of arranging N distinguishable particles in a row, there are $N!$ ways of placing N distinguishable particles in the cells. However, reordering the particles within a cell does not generate a new microstate. Therefore, the number of arrangements that contribute to the multiplicity is $N!$ divided by the number of ways of ordering n_1 particles in the first cell, of ordering n_2 particles in the second cell, ..., and of ordering n_J particles in the Jth cell, that is, by $n_1!n_2!\cdots n_J!$, for a total of $N!/\left(n_1!n_2!\cdots n_J!\right)$ distinct arrangements. This number $N!/\left(n_1!n_2!\cdots n_J!\right)$, clearly a generalization of the binomial coefficient, is called a *multinomial coefficient*.

Therefore, the multiplicity of this occupation number macrostate is

$$\Omega = \frac{N!}{n_1!n_2!\cdots n_J!} \tag{3.24}$$

and the corresponding relative entropy

$$S = c + k \ln \left[\frac{N!}{n_1! n_2! \cdots n_J!} \right]$$

$$= c + k \left[N \ln N - \sum_{j=1}^{J} n_j \ln n_j \right]$$

(3.25)

where in this step we employ Stirling's approximation and in so doing assume that the size of a single phase space cell H^3 is large enough to make the occupation numbers $n_j \gg 1$ for all j.

Maximizing the entropy

We maximize the entropy of this occupation number macrostate with respect to the occupation numbers n_j where these latter are constrained by conservation of particle number

$$N = \sum_{j=1}^{J} n_j$$

(3.26)

and total energy

$$E = \sum_{j=1}^{J} \varepsilon_j n_j.$$

(3.27)

Here ε_j is the energy of a particle when it is in the jth cell of the single-particle phase space. For instance, when the system is an ideal gas the particle energy ε_j is related to its particle momenta $p_{x,j}$, $p_{y,j}$, and $p_{z,j}$ by $\varepsilon_j = \left(p_{x,j}^2 + p_{y,j}^2 + p_{z,j}^2 \right) / 2m$. But other kinds of systems have other relations between particle energy ε_j and particle phase space coordinates $p_{x,j}$, $p_{y,j}$, $p_{z,j}$, x_j, y_j, and z_j.

We use standard methods of calculus to maximize the entropy with respect to the occupation numbers n_i, here considered as continuous variables, subject to the two constraints (3.26) and (3.27). These constraints are introduced into the maximization problem with two Lagrange multipliers α and β. Thus we maximize the entropy of (3.25) by requiring that the partial derivatives of the constrained, normalized, relative entropy

$$N \ln N - \sum_{j=1}^{J} n_j \ln n_j + \alpha \left(N - \sum_{j=1}^{J} n_j \right) + \beta \left(E - \sum_{j=1}^{J} n_j \varepsilon_j \right)$$

(3.28)

with respect to each n_i vanish. The result is $- \ln n_i - 1 - \alpha - \beta \varepsilon_i = 0$ which is solved by

$$n_i = e^{-(1+\alpha)} e^{-\beta \varepsilon_i}$$

(3.29)

where the Lagrange multipliers α and β are, in principle, determined by the constraints (3.26) and (3.27). Since the second partial derivatives of the con-strained relative entropy (3.29) with respect to the occupation number n_j for all j are negative, this procedure locates a relative maximum and not only a stationary value.

Single-particle partition function

We eliminate the Lagrange multiplier α by summing both sides of (3.29) over all cells of the 6-dimensional, single-particle, phase space. This produces

$$N = e^{-(1+\alpha)} \sum_{j=1}^{J} e^{-\beta \varepsilon_j}, \tag{3.30}$$

which, in turn, transforms Eq. (3.29) into the occupation number frequency

$$\frac{n_i}{N} = \frac{e^{-\beta \varepsilon_i}}{\sum_{j=1}^{J} e^{-\beta \varepsilon_j}} \tag{3.31}$$

or, equivalently and more compactly, into

$$\frac{n_i}{N} = \frac{e^{-\beta \varepsilon_i}}{Z_1}, \tag{3.32}$$

where the quantity

$$Z_1 = \sum_{j=1}^{J} e^{-\beta \varepsilon_j} \tag{3.33}$$

is called the *single-particle partition function* and is so symbolized after the German equivalent *Zustandssumme* (state-sum).

The single-particle partition function Z_1 and its generalizations are useful in statistical physics because they simplify many expressions and suggest an efficient order of calculation. For instance, given (3.32) and (3.33) the energy constraint (3.27) becomes

$$\frac{E}{N} = \frac{\sum_{j=1}^{J} \varepsilon_j e^{-\beta \varepsilon_j}}{\sum_{j=1}^{J} e^{-\beta \varepsilon_j}}$$

$$= -\frac{\partial}{\partial \beta} \ln \sum_{j=1}^{J} e^{-\beta \varepsilon_j} \tag{3.34}$$

$$= -\frac{\partial}{\partial \beta} \ln Z_1.$$

This result and (3.32) allow us to express the relative entropy (3.25) as

$$
\begin{aligned}
S(E,V,N) &= c(N) + k \ln\left[N \ln N - \sum_{j=1}^{P} n_j \left(\ln N - \beta \varepsilon_j - \ln Z \right) \right] \\
&= c(N) + k \ln\left[N \ln N - N \ln N + \beta E + N \ln Z_1 \right] \\
&= c(N) + k \left[\beta E + N \ln Z_1 \right],
\end{aligned}
\tag{3.35}
$$

where here we have made explicit the dependence of the entropy S on the extensive fluid variables E, V, and N.

Eliminating β

Ordinarily we would use the energy constraint, (3.27) or (3.34), to eliminate the Lagrange multiplier β in favor of the system energy E. But the algebra of this procedure is difficult if not impossible. Instead, we recall that any time we know how the entropy S of a fluid system depends upon its energy E we know the temperature of that system through $1/T = (\partial S/\partial E)_{V,N}$. To exploit this fact we use the relation (3.35) since it formally expresses the dependence of S on E. Then we find that

$$
\begin{aligned}
\frac{1}{kT} &= \beta + E\left(\frac{\partial \beta}{\partial E} \right)_{V,N} + N\left(\frac{\partial \ln Z_1}{\partial E} \right)_{V,N} \\
&= \beta + \left(\frac{\partial \beta}{\partial E} \right)_{V,N} \left[E + N\left(\frac{\partial \ln Z_1}{\partial \beta} \right) \right].
\end{aligned}
\tag{3.36}
$$

The latter, given the energy constraint (3.34), yields

$$
\beta = \frac{1}{kT}.
\tag{3.37}
$$

Although we could use (3.37) to completely eliminate the Lagrange multiplier β from the entropy in favor of the temperature T, we often find it algebraically convenient to retain β while remembering that $\beta = 1/kT$.

Maxwell–Boltzmann distribution

The result $\beta = 1/kT$ turns the occupation number frequency (3.31) into an expression

$$\frac{n_i}{N} = \frac{e^{-\varepsilon_i/kT}}{\sum_{j=1}^{j} e^{-\varepsilon_j/kT}} \tag{3.38}$$

known as the *Maxwell–Boltzmann distribution*. Recall that in deriving the Maxwell–Boltzmann distribution (3.38) we assumed that the occupation numbers n_j are large compared to 1. When this is so, the ratio n_i/N closely approximates the probability p_i that any one particle will occupy the jth cell of a single-particle, 6-dimensional, phase space. Accordingly, the independent particles of a system infrequently occupy cells with energies large relative to a characteristic thermal energy kT.

Summary

We summarize the occupation number method. The entropy of a system composed of N identical, yet distinguishable, particles is

$$S(E,V,N) = c(N) + k\left[\beta E + N \ln Z_1\right] \tag{3.39}$$

assuming that we can express $\beta\left[=1/kT\right]$ and Z_1 in terms of the extensive variables E, V, and N. Here

$$E = -N\frac{\partial}{\partial\beta}\ln Z_1, \tag{3.40}$$

$$Z_1 = \sum_{j=1}^{J} e^{-\beta\varepsilon_j}, \tag{3.41}$$

and

$$\beta = \frac{1}{kT}. \tag{3.42}$$

Thus, the single-particle partition function (3.41) contains all the physics particular to a given model while expressions (3.39) and (3.40) express algorithms for calculating entropy and energy.

The sum in partition function (3.41) is over all the microstates available to a single particle. These microstates are defined by the different cells of uniform size H^3 that compose the 6-dimensional phase space of a single particle's coordinates: x, y, z, p_x, p_y, and p_z. Equations (3.39)–(3.42) apply to an N-particle system whose particles share a common phase space, independently occupy positions in that phase space, and do not interact except as necessary to maintain system-wide constraints on energy and particle number.

3.5 Ideal classical gas

We apply the occupation number method to an ideal classical gas by determining its single-particle partition function Z_1. In doing so we assume, as did Boltzmann, that the size of a cell in phase space H^3 can be made large enough so that the number of particles in each cell n_j varies only slightly between adjacent cells. Given this assumption the occupation numbers n_j become continuous functions of phase space coordinates, and we can replace the state-sum with an integral

$$\sum_j \rightarrow \int \frac{dx\,dy\,dz\,dp_x\,dp_y\,dp_z}{H^3}. \tag{3.43}$$

Consequently,

$$\sum_j e^{-\beta\varepsilon_j} \rightarrow \frac{1}{H^3}\int dx \int dy \int dz \int dp_x \int dp_y \int dp_z e^{-\beta\varepsilon}, \tag{3.44}$$

where

$$\varepsilon = \frac{p_x^2 + p_y^2 + p_z^2}{2m} \tag{3.45}$$

is the energy of a particle of a monatomic ideal classical gas in terms of its phase space coordinates. And so

$$
\begin{aligned}
\sum_j e^{-\beta\varepsilon_j} &\rightarrow \frac{1}{H^3}\int dx \int dy \int dz \int dp_x \int dp_y \int dp_z\, e^{-\beta\left(p_x^2+p_y^2+p_z^2\right)/2m} \\
&= \frac{V}{H^3}\left[\int_{-\infty}^{\infty} e^{-\beta p^2/2m}\, dp\right]^3 \\
&= \frac{V}{H^3}\left(\frac{2m}{\beta}\right)^{3/2}\left[\int_{-\infty}^{\infty} e^{-s^2}\, ds\right]^3 \\
&= \frac{V}{H^3}\left(\frac{2m\pi}{\beta}\right)^{3/2}
\end{aligned}
\tag{3.46}
$$

where in the last step we have used the identity $\int_{-\infty}^{\infty} e^{-x^2}\, dx = \sqrt{\pi}$. Therefore,

$$Z_1 = \frac{V}{H^3}\left(\frac{2m\pi}{\beta}\right)^{3/2} \tag{3.47}$$

is the single-particle partition function of a monatomic, ideal classical gas. This partition function contains all the physics of this model.

The partition function (3.47) and the Eqs. (3.39)–(3.42) that summarize the occupation number method imply that

$$\begin{aligned}
E &= -N\frac{\partial}{\partial\beta}\ln Z_1 \\
&= -N\frac{\partial}{\partial\beta}\ln\left[\frac{V}{H^3}\left(\frac{2m\pi}{\beta}\right)^{3/2}\right] \\
&= \frac{3}{2}NkT.
\end{aligned} \tag{3.48}$$

Furthermore, the entropy of the classical ideal gas is given by

$$\begin{aligned}
S(E,V,N) &= c(N) + k\left[\beta E + N\ln Z_1\right] \\
&= c(N) + Nk\left\{\frac{3}{2} + \ln\left[\left(\frac{V}{H^3}\right)\left(\frac{2m\pi}{\beta}\right)^{3/2}\right]\right\} \\
&= c(N) + Nk\left\{\frac{3}{2} + \ln\left[\left(\frac{E}{N}\right)^{3/2}\left(\frac{4m\pi}{3H^2}\right)^{3/2}\right]\right\} \\
&= c(N) + k\ln\left[V\left(\frac{E}{N}\right)^{3/2}\left(\frac{4e\pi m}{3H^2}\right)^{3/2}\right]^N.
\end{aligned} \tag{3.49}$$

Note that (3.49) recovers the multiplicity and entropy found in Section 3.2 by discretizing the 6N-dimensional phase space of the ideal gas system. The partial derivatives of the entropy function (3.49), $(\partial S/\partial E)_{V,N}$ $[=1/T]$ and $(\partial S/\partial V)_{E,N}$ $[=P/T]$, yield the equations of state: $E = 3NKT/2$ and $PV = NkT$.

3.6 Ideal classical solid

An ideal classical solid is one whose N particles oscillate independently in simple harmonic motion with a common frequency v_o around a crystalline array of fixed points as illustrated in Figure 3.2. Although these particles do not occupy the same positional space and, therefore, do not occupy the same phase space,

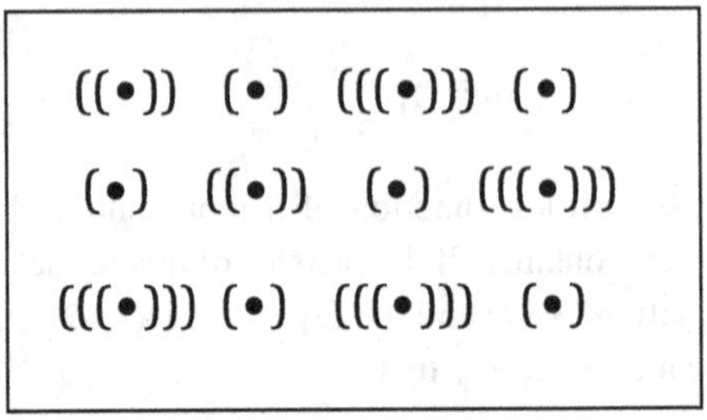

$$((\bullet))\quad(\bullet)\quad(((\bullet)))\quad(\bullet)$$
$$(\bullet)\quad((\bullet))\quad(\bullet)\quad(((\bullet)))$$
$$(((\bullet)))\quad(\bullet)\quad(((\bullet)))\quad(\bullet)$$

Figure 3.2. The classical ideal solid is composed of atoms or molecules that oscillate with simple harmonic motion (same frequency, various amplitudes and phases) around a crystalline array of fixed centers. Although only left–right motion is indicated here, the model includes motion in all three directions.

they do occupy structurally identical phase spaces. For this reason we can use occupations numbers to determine the entropy of the classical ideal solid.

Each particle has three position coordinates, x, y, and z, and three momentum coordinates, p_x, p_y, and p_z. The range of the position coordinates encompasses atomic or molecular sized regions of oscillation rather than the volume V occupied by the entire solid. Also, the energy ε of each particle is the sum of its kinetic and potential energies,

$$\varepsilon = \frac{p_x^2 + p_y^2 + p_z^2}{2m} + \frac{m\omega_o^2}{2}\left(x^2 + y^2 + z^2\right), \tag{3.50}$$

where $\omega_o\left[=2\pi v_o\right]$ is the oscillation frequency in radians per second, v_o is the oscillation frequency in hertz, and $m\omega_o^2$ is the elastic constant. All quantities are the same for all particles and for oscillations in all directions.

Again we replace the state-sum in the partition function

$$Z_1 = \sum_j e^{-\beta\varepsilon_j}$$

with an integral so that

$$\sum_j \rightarrow \frac{1}{H^3}\int dx \int dy \int dz \int dp_x \int dp_y \int dp_z. \tag{3.51}$$

Consequently,

$$\sum_j e^{-\beta\varepsilon_j} \rightarrow \frac{1}{H^3}\int dx \int dy \int dz \int dp_x \int dp_y \int dp_z\, e^{-\beta\varepsilon}$$

$$\rightarrow \frac{1}{H^3}\int dx \int dy \int dz\, e^{-\beta m\omega_o^2\left(x^2+y^2+z^2\right)/2} \int dp_x \int dp_y \int dp_z\, e^{-\beta\left(p_x^2+p_y^2+p_z^2\right)/2m}$$

$$\rightarrow \frac{1}{H^3} \left(\frac{2}{\beta m \omega_o^2} \right)^{3/2} \left(\frac{2m}{\beta} \right)^{3/2} \left[\int_{-\infty}^{\infty} e^{-s^2} ds \right]^6$$

$$= \frac{1}{H^3} \left(\frac{4\pi^2}{\beta^2 \omega_o^2} \right)^{3/2} \tag{3.52}$$

and the single-particle partition function is given by

$$Z_1 = \frac{1}{\left(\beta H v_o \right)^3}. \tag{3.53}$$

Note that in the second step of the equation sequence (3.52) we extend the spatial integration over all space rather than over the system volume V or, more appropriately, over the relatively small volume of a typical sphere of oscillation. This tactic introduces only a small error, because the integrand of the spatial integrations, $\exp\left[-\beta m \omega_o^2 \left(x^2 + y^2 + z^2 \right)/2 \right]$, becomes extremely small when the distance $\sqrt{x^2 + y^2 + z^2}$ from the oscillation center is significantly larger than $\sqrt{2/\left(\beta m \omega_o^2 \right)}$. For this reason, $\sqrt{2/\left(\beta m \omega_o^2 \right)}$ may be taken as the typical radius of the sphere of oscillation. This approximation of the spatial integrations also explains why the single-particle partition function (3.53) of the classical ideal solid is independent of system volume V.

Given (3.53) the energy expression $E = -N \partial \ln Z_1 / \partial \beta$ reduces to $E = 3N/\beta$, that is, to the energy equation of state

$$E = 3NkT. \tag{3.54}$$

Consequently, the heat capacity $C \left[= \left(\partial E / \partial T \right) \right]$ implied by (3.54) is given by

$$= 3Nk, \tag{3.55}$$

a result that reproduces the law of Dulong and Petit according to which the molar specific heat capacity C/n of a solid is always close to the universal constant $3R \left[= 25 \text{ J}/\left(\text{K} \cdot \text{mole} \right) = 6.0 \text{ calorie}/\left(\text{K} \cdot \text{mole} \right) \right]$.

Equipartition theorem

The similarity of the energy equations of state of an N-particle, monatomic, ideal classical gas, $E = 3NkT/2$, and of an N-particle, ideal classical solid, $E = 3NkT$, derives from their similar expressions for particle energy ε, respectively,

$$\varepsilon = \left(p_x^2 + p_y^2 + p_z^2\right)/2m \quad \text{and} \quad \varepsilon = \left(p_x^2 + p_y^2 + p_z^2\right)/2m + m\omega_o^2\left(x^2 + y^2 + z^2\right)/2.$$

Apparently, when the phase space integrations can be extended over all phase space, *each term in the particle energy quadratic in a phase space coordinate contributes NkT/2 to the internal energy of a system of N independent particles* – a statement known as the *equipartition theorem*. When the equipartition theorem applies, a system's internal energy is shared equally among its degrees of freedom. Because the energy of each particle of an N-particle, monatomic, ideal classical gas contains three terms quadratic in the momenta, its internal energy E is $3 \times N \times kT/2$. Because each oscillator of an ideal classical solid contains six terms quadratic in its coordinates, its internal energy E is $6 \times N \times kT/2$.

The entropy of an ideal classical solid

Finally, we produce an expression for the entropy of an ideal classical solid in terms of its macrostate internal energy E and particle number N. According to the occupation number method, $S = c(N) + k(\beta E + N \ln Z_1)$ where $\beta = 1/kT$, and, according to our model of the classical, ideal solid $E = 3NkT$ and $Z_1 = 1/(\beta Hv_o)^3$. Together these yield

$$S(E,N) = c(N) + 3Nk\left[1 + \ln\left(\frac{E}{3NHv_o}\right)\right]. \tag{3.56}$$

Because this entropy does not depend upon the volume V, the pressure of an classical ideal solid vanishes – an unrealistic consequence of this overly simple model.

Example 3.2 Average kinetic energy

Problem: What is the average kinetic energy of a particle of classical, monatomic, ideal gas? And what is the root mean squared speed of a particle of this gas?

    Solution: Since all the energy of a classical, monatomic, ideal gas is contained in the kinetic energy of its particles, the answers to these questions follow almost immediately from the monatomic, ideal classical gas equation of state $E=3NkT/2$. The average kinetic energy per particle, $E/N=3kT/2$, is a function of the thermodynamic temperature T alone. This proportionality between the average kinetic energy and the thermodynamic temperature obtains for systems that observe the equipartition theorem.

    If m is the mass of each gas particle, then the root mean squared velocity of the particles $\sqrt{\langle v^2 \rangle}$ is determined by

$$\frac{m\langle v^2 \rangle}{2} = \frac{3kT}{2}.$$

Therefore,

$$\sqrt{\langle v^2 \rangle} = \sqrt{\frac{3kT}{m}}.$$

The hotter the gas, the faster its particles; the more massive the particle, the more slowly it moves.

3.7 Boltzmann's tomb

It is perhaps surprising that the classical way of doing statistical mechanics has so little to do with mechanics *per se*, that is, with the science of motion. Nowhere do the methods of classical statistical mechanics require $F = ma$. Nowhere do they require solutions to equations of motion. Particles move, but move probabilistically according to certain rules: the fundamental postulate, conservation of energy, and conservation of particle number. That Boltzmann fashioned these ideas into a statistical method is, in retrospect, prescient. For, as we shall see, Boltzmann's approach to entropy is almost perfectly congruent with the quantum physics that emerged after his death.

Recall that the relation of relative entropy to macrostate multiplicity, $S = c + k \ln \Omega$, follows from assuming that: (1) the entropy of an isolated system is a function $S(\Omega)$ of macrostate multiplicity Ω; (2) the entropies of isolated subsystems are additive; and (3) their macrostate multiplicities are statistically independent. The most direct way of calculating the multiplicity Ω is, as in Sections 3.1 and 3.2, to discretize the phase space available to a system of given energy E, volume V, and number N of distinguishable particles and count the number of cells.

In practice Boltzmann preferred the relatively indirect, yet powerful, occupation number method of determining the entropy of a system composed of N independent, distinguishable particles. According to this method we (1) distribute the particles in a one-particle, 6-dimensional phase space that is divided into J cells; (2) define a macrostate in which the N particles of the system are partitioned into the numbers of particles, $n_1, n_2, \ldots, n_J$, that occupy its J cells; (3) form the multiplicity Ω of this macrostate $N!/(n_1!n_2!\ldots n_J!)$; and (4) determine the macrostate that maximizes its entropy $S\left[= c + k \ln \Omega\right]$ subject to the constraints $N = n_1 + n_2 + \lambda + n_j$ and $E = n_1\varepsilon_1 + n_2\varepsilon_2 + \cdots + n_j\varepsilon_j$. Both the direct and the occupation number methods produce a relative entropy rather than an absolute entropy.

What then is one to make of the formula for absolute entropy, $S = k \log W$, found on Boltzmann's tomb in the Vienna *Zentralfriedhof* (Central Cemetery)? Evidently, those who dedicated Boltzmann's tomb in 1933 believed that $S = k \log W$ encapsulated Boltzmann's contribution to physics. But this equation appears nowhere in Boltzmann's writings, originates with Max Planck, and incorporates physics of which Boltzmann could not have been aware. The last of these three statements may be controversial and, consequently, needs justification.

But first, what does $S = k \log W$ mean? The convention that S stands for entropy had been established by Clausius in 1865 and k was Planck's choice for what we now call Boltzmann's constant (probably after the German word *Konstant*). But the letter W, taken from the German word *Wahrscheinlichkeit* for *probability*, is problematic. Certainly the word *probability* misleads. For probabilities are usually normalized to lie between 0 and 1 inclusive. And if $0 \leq W \leq 1$, the equation $S = k \log W$ leads to negative entropy – contrary to Clausius's convention and our expectation. To be sure, Planck envisioned other ways of normalizing probabilities and called W not the *probability* but rather the *thermodynamic probability*. Others have referred to W as the *thermodynamic* or *statistical weight*. In actual practice, Boltzmann always used W as a part of the occupation number method and always replaced W by the multinomial coefficient $N!/(n_1!n_2!.\lambda n_j!)$. Thus W appears to be equivalent to what we call the multiplicity Ω of an occupation number macrostate of a system of N distinguishable particles.

Given these identifications, Boltzmann's epitaph, $S = k \log W$, is equivalent to $S = k \ln \Omega$. And so we return to the question: What is the physics required to turn the equation describing the entropy of a classical system, $S = c + k \ln \Omega$, into the equation inscribed on Boltzmann's tomb? As we shall see in Section 4.5, this missing physics is the third law of thermodynamics – a law that allows one to adopt a conventional value for the entropy in the $T \to 0$ limit. But Walther Nernst (1884–1941) first articulated a version of the third law in 1906 – too late for Boltzmann to exploit before his death in September of that year. And neither the ideal gas nor the ideal solid models that Boltzmann studied are third law compliant. Evidently the equation $S = k \log W$ summarizes not only Boltzmann's direct contribution to physics but also his service in laying a foundation upon which a new quantum statistical physics, then only emerging at the time of his death, could be built.

Problems

3.1 Room temperature density

Determine the number of particles in 1 cm³ of ideal gas when $T = 320$ K and the pressure is 1 atmosphere. (Consult Appendix 1: Physical constants and standard definitions.)

3.2 Van der Waals equations of state

The goal of much of statistical mechanics is to produce meaningful partition functions. But suppose someone gives you the ready-made single-particle partition function

$$Z_1 = (V - Nb)\left(\frac{2\pi}{m\beta}\right)^{3/2} e^{\beta a N / V},$$

where a and b are constants that characterize a gas system and $\beta = 1/kT$. Find (a) the corresponding energy equation of state and (b) the entropy function $S(E, V, N)$. (Note that β is a function of the extensive variables E, V, and N.) From the entropy function $S(E,V,N)$ find (c) the energy equation of state again and (d) the pressure equation of state.

3.3 Ideal gas of diatomic molecules

The single-particle partition function of an ideal gas of diatomic molecules is

$$Z_1 = V\left(\frac{2\pi}{m\beta}\right)^{5/2},$$

where $\beta = 1/kT$. Find (a) the corresponding energy equation of state and the pressure equations of state. (b) Use the equipartition theorem to make a mean-ingful statement about the number of quadratic terms in the energy of the diatomic molecules in this gas.

3.4 Mixing of ideal gases

Two containers of ideal gas, one of N_2 and another of O_2 molecules, have the same volume V, pressure P, and temperature T. A valve connecting the two containers is opened and the resulting mixture comes to equilibrium. Assume the container walls are adiabatic. What is the resulting increase ΔS in the entropy of this two-gas system? Use Eq. (3.49).

3.5 Extensivity of ideal solid

Impose the extensivity condition $S(\lambda E, \lambda N) = \lambda S(E, N)$ on the entropy of an ideal classical solid given in (3.56). (Note: In much the same way that there is a family of extensive entropies of an ideal classical gas, there is a family of extensive entropies of the ideal classical solid.)

3.6 Room temperature speed of N_2 molecules

Use the data given in Appendix I, Physical Constants and Standard Definitions, to determine the root mean squared speed in meters per second of an N_2 molecule at room temperature ($T = 300$ K). The mass of a nitrogen molecule is 28.0 u.

4

Entropy of quantized systems

4.1 Quantum conditions

Max Planck (1858–1947) initiated the era of quantum physics on December 14, 1900 at a meeting of the German Physical Society by announcing a derivation of the spectral energy density of blackbody radiation that was consistent with the most recent data. Planck's intention had been to apply Maxwell's equations and Boltzmann's statistical mechanics to electromagnetic radiation within a cavity whose walls were at a uniform temperature T, that is, to equilibrium or *blackbody radiation*. But along the way something unexpected happened: the arbitrary quantity H that regulates the size of the phase space cells remained in expressions of quantities that could be measured. The data available to Planck suggested that H has a definite, fixed value, henceforth denoted h, that is close to the modern value,

$$h = 6.63 \times 10^{-34} \frac{\text{m}^2 \text{ kg}}{\text{s}}, \tag{4.1}$$

of this universal constant now known as Planck's constant.

Historians of science still debate the extent to which Planck was aware in 1900 that his derivation depended upon revolutionary ideas. However, by 1908 Planck was well aware and by 1925 physical scientists understood that quantum physics includes the following conditions: (1) phase space cells have a definite size determined by Planck's constant h; (2) the energy, momenta, and other dynamical properties of an isolated system are quantized; and (3) identical particles are, for the purpose of determining multiplicity, indistinguishable from one another.

While classically inexplicable, these conditions emerge naturally from the wave mechanics that succeeded the old quantum theory of 1900–1925. And while Clausius's thermodynamic conception of entropy and Boltzmann's

framework for relating the entropy of a system to its macrostate multiplicity
survived the quantum revolution, the way we determine macrostate multiplicities has evolved.

The classical way of doing statistical mechanics remains coherent and
appropriate for working in, for instance, that regime in which the ideal gas
equations of state are valid. But the classical way of doing statistical mechanics is also limited in its application and when applied outside of its appropriate
regime makes predictions that are not in accord with measurements.

A minimal application of the quantum conditions is simply a reinterpre-
tation of the extensive entropy of an ideal gas (3.22) with Planck's constant h
replacing the arbitrary unit of action H. The result

$$S(E,V) = Nk \left\{ \frac{5}{2} + \ln \left[\left(\frac{V}{N} \right) \left(\frac{E}{N} \right)^{3/2} \left(\frac{4\pi m}{3h^2} \right)^{3/2} \right] \right\} \tag{4.2}$$

is called the Sackur–Tetrode entropy after the two men who developed
it independently in 1912. Recall that Boltzmann invented the quantity H
$\left[\equiv \delta s \, \delta x \right]$ in order to discretize phase space, but did not need to give H a definite value. Quantum physics, however, needs to and does assign H the def-
inite value $h \left[= 6.63 \times 10^{-34} \text{ m}^2 \text{ kg/s} \right]$. Using Planck's constant to discretize
phase space produces in the Sackur–Tetrode equation (4.2) an expression
for the absolute, extensive entropy of a quantized ideal, monatomic gas in
the semi-classical regime which, as we shall see, is the low density, high
temperature regime.

4.2 Quantized harmonic oscillators

The quantum harmonic oscillator exploits two quantum conditions: determinate phase space cells and quantized energy.

Consider an array of N identical, simple harmonic oscillators that vibrate
independently around fixed centers with frequency $\omega_o \left[= 2\pi v_o \right]$ where v_o is in
hertz. Although identical, these oscillators are distinguishable by virtue of the
distinct positions of their centers of oscillation. The force constant of the linear
force that returns the oscillators to their equilibrium positions is $m\omega_o^2$ where m
is the oscillator mass.

When these oscillators are one-dimensional, the state of each can be represented as a point in two-dimensional, single-particle, x–p phase space as illustrated in Figures 4.1a and 4.1b.

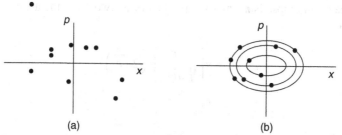

Figure 4.1 Oscillator positions in two-dimensional, single-particle, x–p phase space. (a) Oscillators randomly placed. (b) Oscillators placed on constant-energy ellipses with each ellipse separated by energy hv_o.

Oscillator energy quantized

Figure 4.1a shows oscillators that seem randomly positioned in phase space. However, if we could observe their positions in sufficiently fine scale, we would see the oscillators naturally occupying positions on equal-energy ellipses as in Figure 4.1b For since the energy ε of a single, one-dimensional oscillator is related to its position x and momentum p by

$$\frac{m\omega_o^2 x^2}{2} + \frac{p^2}{2m} = \varepsilon, \tag{4.3}$$

oscillators that lie on the ellipse in phase space described by

$$\frac{x^2}{\left(2\varepsilon/m\omega_o^2\right)} + \frac{p^2}{\left(2m\varepsilon\right)} = 1 \tag{4.4}$$

have the same energy ε. Furthermore, oscillators on the smallest ellipse have energy $\varepsilon_0 = \frac{1}{2}hv_o$, oscillators on the second smallest ellipse have energy $\varepsilon_1 = \left(1 + \frac{1}{2}\right)hv_o$, oscillators on the third smallest ellipse have energy $\varepsilon_2 = \left(2 + \frac{1}{2}\right)hv_o$, and so on. Apparently, oscillator energy is quantized according to the rule

$$\varepsilon_j = \left(j + \frac{1}{2}\right)hv_o, \tag{4.5}$$

where $j = 0, 1, 2, \dots$.

Oscillator phase space quantized

One consequence of quantizing the oscillator energy in this way is that the area surrounding each ellipse in oscillator phase space is itself quantized. For since the area of an ellipse is π times the product of its two radii, the area A_j of the jth ellipse is

$$A_j = \pi \left(\sqrt{\frac{2\varepsilon_j}{m\omega_o^2}} \right) \left(\sqrt{2m\varepsilon_j} \right)$$

$$= \frac{\varepsilon_j}{v_o} \tag{4.6}$$

$$= h \left(j + \frac{1}{2} \right),$$

where in the second step of this equation sequence we have used (4.5). Therefore,

$$A_{j+1} - A_j = h, \tag{4.7}$$

that is, the area between the $j+1$ and j ellipses in the phase space of the quantized harmonic oscillator is Planck's constant h.

If the area of the smallest phase space cell of a one-dimensional system is h, what is its shape? In particular, why shouldn't the shape of the smallest cells in oscillator phase space be that of elliptical annuli enclosing each equal-energy ellipse? After all, there can be no universal cell shape. For what would that cell shape be: rectangles, circles, wedges? Apparently each system has a cell shape appropriate to its dynamical structure. And, if the cells of oscillator phase space are elliptical annuli that enclose equal-energy ellipses, each such annulus has an area h. In this case, points that lie on one ellipse lie within one cell.

Oscillator entropy

Since the quantized oscillators are distinguishable one from another by virtue of their separate positions, the number of oscillators on the jth ellipse is the occupation number n_j of the jth cell and Boltzmann's occupation number method, as summarized in Eqs. (3.39)–(3.42), applies. Furthermore, we need no longer assume that the occupation numbers n_j are continuous functions of x and p. We can do the required sum exactly! This is fortunate since we can no longer assume that the single-particle phase space cells contain many particles. In particular, given that the microstate index j goes from 0 to ∞,

$$Z_1 = \sum_{j=0}^{\infty} e^{-\beta\varepsilon_j}$$

$$= \sum_{j=0}^{\infty} e^{-\beta h v_o (j+1/2)}$$

$$= e^{-\beta h v_o/2} \left[1 + e^{-\beta h v_o} + e^{-2\beta h v_o} + e^{-3\beta h v_o} + \cdots \right]$$

$$= e^{-\beta h v_o/2}\left[1+\left(e^{-\beta h v_o}\right)^1+\left(e^{-\beta h v_o}\right)^2+\left(e^{-\beta h v_o}\right)^3+\cdots\right]$$

$$= \frac{e^{-\beta h v_o/2}}{1-e^{-\beta h v_o}}. \tag{4.8}$$

Since $\beta = 1/kT$ and, according to the occupation number method, $n_j/N = e^{-\varepsilon_j/kT}/Z_1$, the occupation number frequencies of a one-dimensional oscillator are given by

$$\frac{n_j}{N} = e^{-jhv_o/kT}\left(1-e^{-hv_o/kT}\right), \tag{4.9}$$

where $j = 0,1,2,\dots$.

The energy of a system of N one-dimensional oscillators is

$$E = -N\frac{\partial}{\partial\beta}\ln Z_1$$

$$= Nhv_o\left[\frac{1}{2}+\frac{1}{\left(e^{hv_o/kT}-1\right)}\right]. \tag{4.10}$$

Solving (4.10) for hv_o/kT in terms of the normalized system energy E/Nhv_o, we find, after some algebra, that

$$\frac{hv_o}{kT} = \ln\left(\frac{\dfrac{E}{Nhv_o}+\dfrac{1}{2}}{\dfrac{E}{Nhv_o}-\dfrac{1}{2}}\right). \tag{4.11}$$

This relation (4.11), the energy equation of state (4.10), and the partition function (4.8) allow us to determine the entropy $S(E,N)$ in terms of the system's extensive thermodynamic variables. We find that

$$S(E,N) = c(N)+k\left[\beta E + N\ln Z_1\right]$$

$$= c(N)+Nk\left\{\left(\frac{E}{Nhv_o}+\frac{1}{2}\right)\ln\left(\frac{E}{Nhv_o}+\frac{1}{2}\right)\right.$$

$$\left.-\left(\frac{E}{Nhv_o}-\frac{1}{2}\right)\ln\left(\frac{E}{Nhv_o}-\frac{1}{2}\right)\right\}. \tag{4.12}$$

For the most efficient way of achieving this result, see Problem 4.1.

Three-dimensional oscillators

Statistically $3N$ one-dimensional oscillators are equivalent to N three-dimensional oscillators when all have the same frequency v_o. Therefore, the energy of N three-dimensional oscillators is given by (4.10) with N replaced by $3N$, that is, by

$$E = 3Nhv_o \left[\frac{1}{2} + \frac{1}{\left(e^{hv_o/kT} - 1\right)} \right], \tag{4.13}$$

and the entropy of a system of N three-dimensional oscillators is given by (4.12) with N replaced by $3N$, that is, by

$$S(E,N) = c(N) + 3Nk \left\{ \left(\frac{E}{3Nhv_o} + \frac{1}{2} \right) \ln \left(\frac{E}{3Nhv_o} + \frac{1}{2} \right) \right.$$
$$\left. - \left(\frac{E}{3Nhv_o} - \frac{1}{2} \right) \ln \left(\frac{E}{3Nhv_o} - \frac{1}{2} \right) \right\}, \tag{4.14}$$

where we have redefined $c(N)$. Albert Einstein (1879–1955) first derived and applied these results in 1907.

Example 4.1 Correspondence principle

Problem: What is the relation between the classical entropy and the quantum entropy of a system of N three-dimensional, linear, harmonic oscillators?

Solution: The energy (4.13) and the entropy (4.14) of a solid composed of quantized oscillators should reduce to their classical counterparts in the appropriate limit. In particular, the quantum entropy (4.14) should reduce to the classical entropy of an ideal solid (3.56) with H replaced by h,

$$S(E,N) = c(N) + 3Nk \left[1 + \ln \left(\frac{E}{3Nhv_o} \right) \right],$$

of an ideal solid. Likewise the quantum energy (4.13) should reduce to the classical energy (3.54) of an ideal solid,

$$E = 3NkT.$$

Indeed, these reductions obtain in the limit of high energy per oscillator E/N relative to the unit hv_o in which energy is quantized, that is, for $E/N \gg hv_o$ or, equivalently, for high thermal energy kT relative to hv_o, that is, for $kT \gg hv_o$. Such is one use of the *correspondence principle* according to which there is always a limit, the so-called *semi-classical limit*, in which quantum expressions reduce to classical ones with the arbitrary H replaced with Planck's constant h.

4.3 Einstein solid

Albert Einstein was probably the first, in 1907, to deliberately apply the quantum conditions to a physical model – to what has become known as the *Einstein solid*. An Einstein solid is a crystalline array of atoms or molecules each one of which oscillates simple harmonically in three dimensions with a common frequency v_0. Einstein derived the energy (4.13) and the entropy (4.14) of a system of N such three-dimensional oscillators.

Einstein was primarily concerned with the heat capacity $C = dE/dT$ of a solid, that is, with

$$C = 3nN_A k \frac{\left(hv_o/kT\right)^2 e^{hv_o/kT}}{\left(e^{hv_o/kT} - 1\right)^2}, \tag{4.15}$$

where here n is the number of moles and N_A is Avogadro's number. The high temperature, semi-classical regime, $kT \gg hv_o$, of this heat capacity recovers the law of Dulong and Petit, $C = 3nN_A k \left[= 3nR\right]$. However, as $kT/hv_o \to 0$ the molar specific heat capacity $C/n \to 0$. In 1907 physical scientists were just becoming aware of molar specific heat capacities that drop below their Dulong–Petit value at relatively low temperatures.

Einstein compared the dependence of C/n on T described by (4.15) to data available on the specific heat of diamond in one of only three graphical comparisons of theory and experiment he ever published. In doing so he used the quantity hv_o/k, now known as the *Einstein temperature*, as a parameter with which to fit the data and found that $hv_o/k = 1325$ K worked well for diamond. I have reproduced Einstein's graph with his data converted to SI units in Figure 4.2.

The Einstein solid is a rough model that in its details must be inaccurate. After all, because neighboring molecules exert the forces that restore a molecule to its equilibrium position, neighboring molecules do not oscillate independently as is assumed in deriving the Einstein solid. The resulting collective motion generates a whole spectrum of oscillations, not just a single frequency. Peter Debye (1884–1966) incorporated these collective oscillations into a statistical model of solids that better fits specific heat data. But

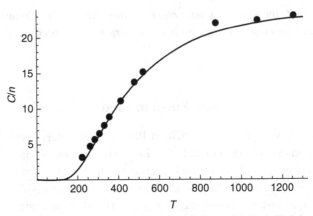

Figure 4.2 Molar specific heat capacity C/n of diamond in joule/(kelvin mole) versus thermodynamic temperature T in kelvin. Filled circles: Measurements available to Einstein in 1907. Curve: Einstein's prediction (4.15) with $hv_o/k = 1325$ K.

it was Einstein's 1907 paper that initiated the modern study of condensed matter physics.

4.4 Phonons

Because the energy of a simple harmonic oscillator is quantized in units of energy hv_o where v_o is the natural frequency of the oscillator, the energy of a system composed of such oscillators is also quantized. In particular, the normalized system energy in excess of its minimum, zero-temperature or *ground state*, value, that is, $(E - 3Nhv_o/2)/hv_o$ is an integer. In the context of an Einstein solid hv_o-sized bundles or quanta of energy are called *phonons*. Thus, energy flows from one place to another in the solid as phonons diffuse from one place to another.

This way of thinking about the energy of an Einstein solid suggests an alternative derivation of its entropy that uses each of three quantum conditions: (1) determinate phase space cells, (2) quantized energy, and (3) the quantum indistinguishability of phonons. Imagine that each oscillator degree of freedom is a compartment that can hold any number of phonons. The Einstein solid consists of $3N$ such compartments. Together these compartments hold phonons.

○ ○ | ○ | ○ ○ ○ ○ | |

Figure 4.3 An arrangement of seven phonons in five ordered compartments. The symbol ○ stands for a phonon and | for a divider between the compartments that hold the phonons. From left to right these compartments hold 2, 1, 4, 0 and 0 phonons.

$$n = \frac{\left(E - 3Nh v_o / 2\right)}{h v_o} \tag{4.16}$$

The number of distinct ways of distributing n indistinguishable phonons among $3N$ compartments is the system multiplicity Ω. To motivate our calculation of the multiplicity let the symbol ○ stand for a phonon and | for a divider between compartments. One distribution of seven phonons into five compartments is shown in Figure 4.3. The order of compartments from left to right correlates with the order of oscillator sites and degrees of freedom. Here the first compartment contains two phonons, the second one phonon, the third four, and the fourth and fifth no phonons.

Thus, the number of distinct ways n indistinguishable phonons may be put into $3N$ compartments is $(n + 3N - 1)! / \left[n!(3N - 1)! \right]$ – the binomial coefficients again. This counting procedure naturally incorporates the conservation of certain quantities: the system energy $E\left[= h v_o \left(n + 3N/2\right) \right]$ and the number of oscillators N. Therefore, the multiplicity of a macrostate consisting of n phonons and N three-dimensional oscillators is

$$\Omega = \frac{(n + 3N - 1)!}{n!(3N - 1)!}, \tag{4.17}$$

and its entropy is

$$\begin{aligned}
S(E, N) &= c(N) + k \ln \left[\frac{(n + 3N - 1)!}{n!(3N - 1)!} \right] \\
&= c(N) + k \left[(n + 3N) \ln(n + 3N) - n \ln n - (3N) \ln(3N) \right] \\
&= c(N) + k \left[\left(\frac{E}{h v_o} + \frac{3N}{2} \right) \ln \left(\frac{E}{h v_o} + \frac{3N}{2} \right) \right. \\
&\quad \left. - \left(\frac{E}{h v_o} - \frac{3N}{2} \right) \ln \left(\frac{E}{h v_o} - \frac{3N}{2} \right) - (3N) \ln(3N) \right]
\end{aligned}$$

$$
\begin{aligned}
= c(N)+3Nk &\left[\left(\frac{E}{3Nhv_o}+\frac{1}{2}\right)\ln\left(\frac{E}{hv_o}+\frac{3N}{2}\right)\right.\\
&\left.-\left(\frac{E}{3Nhv_o}-\frac{1}{2}\right)\ln\left(\frac{E}{hv_o}-\frac{3N}{2}\right)-\ln(3N)\right]\\
= c(N)+3Nk &\left[\left(\frac{E}{3Nhv_o}+\frac{1}{2}\right)\ln\left(\frac{E}{3Nhv_o}+\frac{1}{2}\right)\right.\\
&\left.-\left(\frac{E}{3Nhv_o}-\frac{1}{2}\right)\ln\left(\frac{E}{3Nhv_o}-\frac{1}{2}\right)\right],
\end{aligned}
$$

$$(4.18)$$

where the first step in this sequence of equations follows from $3N \gg 1$ and $n \gg 1$ and the second from (4.16). The fourth step redefines $c(N)$. This entropy (4.18) is the same as that (4.14) determined in Section 4.2 via occupation numbers.

The method of phonons determines the entropy of a system directly in terms of the thermodynamic variables – here energy E and oscillator number N – without maximizing the constrained entropy as a function of occupation number macrostate. When such direct methods work, they often work well.

Indistinguishable particles and distinguishable places

The direct method makes explicit the indistinguishability of identical phonons. After all, phonons are simply quanta of energy. Still one might object that phonons can become distinct from one another in much the same way that identical classical particles are so distinguished: by virtue of being confined to widely separated regions of phase space. But this could only be the case if there were many more oscillators than phonons, that is, $N \gg n$, a regime that does, indeed, coincide with the classical limit. More generally, identical phonons are indistinguishable from each other. Thus, when two phonons are imagined to exchange positions a new microstate is not generated, just as when two dividers between the compartments that hold the phonons are exchanged a new microstate is not generated.

However, the compartments themselves, that is, the oscillators, remain distinguishable from each other – in Figure 4.3 by their very order. In general, the oscillators within a crystalline array are distinguished by the different places they occupy. Which identical objects are indistinguishable and which are not is a matter of context. In general those identical objects that permanently occupy distinct places in phase space are distinguishable from one another, while those identical objects that may occupy a common set of positions in phase space are indistinguishable.

4.5 Third law

We will continue to use the direct method of counting microstates and determining the multiplicity Ω and entropy S of other systems, but first we examine the consequences of applying the third law of thermodynamics. Recall the third law of thermodynamics: *In the limit of vanishing thermodynamic temperature the entropy of a system is independent of its remaining thermodynamic variables.* Max Planck observed that one is free to add to this empirical content a conventionally chosen value for this limiting entropy.

Neither the monatomic, ideal classical gas nor the ideal classical solid observes the third law. After all, since the energy equation of state of each is such that $E \propto T$, then $E \to 0$ as $T \to 0$. And given that the entropy S of each is a linear function of $\ln E$, the entropy of each diverges to negative infinity as $T \to 0$.

However, the Einstein solid does observe the third law. For, as $T \to 0$ the energy of the Einstein solid, given in (4.13), has the limit $E \to 3Nh\nu_o/2$. Therefore, the entropy of the Einstein solid, given in (4.14) or (4.18), has the limit $S(E) \to c(N)$. Since, in this case, we do not need the "constant" $c(N)$ in order to make the entropy extensive, we may, indeed, set $c(N)$ equal to zero, in order that the $T \to 0$ limit of the entropy becomes independent of all thermodynamic variables. It may seem odd that a law of thermodynamics, usually thought of as a classical subject, depends so intimately on quantum physics for its observance. Yet, such is the case.

The third law allows us to reinterpret and to simplify the relative entropy $S\ [=c+k\ln\Omega]$. First, observe that the relative entropy may be written in a form

$$S(\Omega) = S(1) + k\ln\Omega \tag{4.19}$$

in which the constant c has been replaced by the entropy $S(1)$ of a macrostate composed of one microstate. Then, observe that $\Omega \to \Omega_{T=0}$ as $T \to 0$, and therefore that the entropy $S \to S(\Omega_{T=0})$ as $T \to 0$. Applying this limit to the relative entropy (4.19) produces

$$S(\Omega_{T=0}) = S(1) + k\ln\Omega_{T=0}. \tag{4.20}$$

Using (4.20) to eliminate $S(1)$ from the relative entropy (4.19) yields

$$S(\Omega) = S(\Omega_{T=0}) + k\ln\frac{\Omega}{\Omega_{T=0}}. \tag{4.21}$$

According to the third law $S(\Omega_{T=0})$, and consequently $\Omega_{T\to0}$, must be independent of the thermodynamic variables that define a macrostate. This statement incorporates the empirical content of the third law but not yet Planck's convention.

Planck's convention

Planck's convention, that we are free to impose a convenient value on the ground state entropy $S(\Omega_{T=0})$, allows us to go one step further. A commonly adopted convention is to choose

$$S(\Omega_{T=0}) = k \ln \Omega_{T=0} \qquad (4.22)$$

so that $S(1) = 0$ and the relative entropy (4.21) reduces to a version of the absolute or third law entropy

$$S(\Omega) = k \ln \Omega \qquad (4.23)$$

of an isolated system. Alternatively, we could choose $S(\Omega_{T=0}) = 0$ so that (4.21) reduces to $S(\Omega) = k \ln(\Omega/\Omega_{T=0})$. The first choice, (4.22) resulting in (4.23), is traditional. We follow this tradition by adopting $S = k \ln \Omega$ for the absolute entropy of a fully quantized and, subsequently, extensive, third law compliant system.

In actual fact the two conventions are rarely distinct. Not only do fully quantized models of thermodynamic systems observe the third law of thermodynamics, but fully quantized models also usually have only one microstate at absolute zero, that is, usually $\Omega_{T=0} = 1$. In this case, $S(\Omega_{T=0}) = 0$ as well as $S(1) = 0$, and the two conventions collapse into one. However, useful models have been constructed of systems that are degenerate at $T = 0$, that is, for which $\Omega_{T=0} > 1$. In this case, the absolute entropy $S[= k \ln \Omega]$ of a system with a degenerate ground state does not vanish in the limit of vanishing thermodynamic temperature T.

Constructing the entropy

Nonetheless, many texts baldly assert $S = k \ln \Omega$ as a definition of the statistical entropy of an isolated system. After all, this compact equation, first proposed by the great Max Planck, contains much physics. And that its equivalent appears on Boltzmann's tomb adds to its appeal. But simply to assert $S = k \ln \Omega$ makes sense only for fully quantized systems that are third law compliant. Applying $S = k \ln \Omega$ to a classical system such as the ideal gas invites paradox.

Our strategy, beginning in Chapter 2, has been to require that the statistical entropy be consistent with the first and second laws of thermodynamics and to construct its form out of desired properties: dependence on multiplicity Ω, additivity, and the independence of isolated subsystems. This construction produces an entropy $S \left[= S(1)+k\ln\Omega\right]$ relative to the entropy $S(1)$ of a macrostate consisting of a single microstate and relative to an arbitrary phase space unit H.

The quantum conditions require that Planck's constant h replace the arbitrary unit H. We also find that the third law of thermodynamics allows one to eliminate the entropy constant $c \left[= S(1)\right]$. Together these developments turn the relative entropy of classical thermodynamics into the absolute entropy of quantized, third law compliant systems. The option of using the absolute entropy $S \left[= k\ln\Omega\right]$ arises only for quantized models. For only quantized models remain accurate at low temperatures and, for this reason, only quantized models are third law compliant.

Example 4.2 Shottky defect

Problem: Determine the entropy $S(E,n)$ of a crystal consisting of n lattice sites N of which are filled with molecules or ions and $n - N$ of which are empty. Suppose the N particles occupying lattice sites in the crystal contribute no energy to the system while each of the $n - N$ empty lattice sites adds energy ε to the system. Express the entropy S as a function of temperature T and determine whether or not this system is third law compliant.

Solution: First we find the absolute entropy $S = k\ln\Omega$ consistent with a given number of lattice sites n and particles N. The number of distinct ways of choosing $n - N$ empty sites from among n possibilities is the binomial coefficient, that is, "n choose $n - N$,"

$$\Omega = \frac{n!}{(n-N)!N!},$$

where the number of empty sites $n - N$ is related to the energy of the system E by

$$E = (n-N)\varepsilon.$$

Therefore, the entropy in terms of E and n is

$$\frac{S(E,n)}{k} = n\ln n - (n-N)\ln(n-N) - N\ln N$$

$$= n\ln n - \left(\frac{E}{\varepsilon}\right)\ln\left(\frac{E}{\varepsilon}\right) - \left(n - \frac{E}{\varepsilon}\right)\ln\left(n - \frac{E}{\varepsilon}\right).$$

The relation $1/T = (\partial S/\partial E)_n$ produces the equation of state

$$E = \frac{n\varepsilon}{1 + e^{\varepsilon/kT}}.$$

Using this result to eliminate the energy E in favor of the temperature T in the expression for entropy S produces

$$S(T) = nk\left[\ln\left(1 + e^{\varepsilon/kT}\right) - \frac{\left(\varepsilon/kT\right)e^{\varepsilon/kT}}{1 + e^{\varepsilon/kT}}\right].$$

Finally, as $T \to 0$, we see that $\varepsilon/kT \to \infty$ and $S \to 0$. Therefore, this system is third law compliant.

4.6 Paramagnetism

A paramagnet is composed of nuclear, atomic, or molecular magnetic dipoles, essentially small magnets whose dipole moments tend to align with an externally applied magnetic field. Here we consider an especially simple kind of paramagnet composed of spin-½ magnetic dipole moments fixed in a crystal lattice. The quantum mechanics of these spin-½ dipoles requires that each dipole moment m_B is either parallel or antiparallel to the applied field B_o.

When parallel to the applied field B_o each magnetic dipole decreases the system energy by $m_B B_o$. Alternatively, when antiparallel each magnetic dipole adds $m_B B_o$ to the system energy. If n_+ is the number of parallel and n_- the number of antiparallel magnetic dipoles and all N lattice sites are filled with spin-½ paramagnets, then

$$N = n_+ + n_-, \tag{4.24}$$

and

$$\begin{aligned} E &= (n_- - n_+)m_B B_o \\ &= (N - 2n_+)m_B B_o. \end{aligned} \tag{4.25}$$

Consequently,

$$n_+ = \frac{N}{2} - \frac{E}{2m_B B_o} \tag{4.26a}$$

and

$$n_- = \frac{N}{2} + \frac{E}{2m_B B_o}. \tag{4.26b}$$

Thus, macrostates for which $E = 0$ are macrostates in which there are equal number of parallel and antiparallel dipoles. When the energy is near its minimum value $E = -Nm_B B_o$, most of the dipoles are parallel to the applied field so that $n_+ \approx N$ and $n_- \approx 0$. And when the energy is near its maximum value $E = Nm_B B_o$, most of the dipoles are antiparallel to the applied field so that $n_+ \approx 0$ and $n_- \approx N$.

Entropy

The absolute entropy $S\ [= k \ln \Omega]$ of this system depends on the number of ways n_+ dipoles out of N can be chosen, that is, on N choose n_+. Thus,

$$\Omega = \frac{N!}{n_+!(N-n_+)!} \tag{4.27}$$

and so

$$
\begin{aligned}
S &= k \ln \Omega \\
&= k \ln \left[\frac{N!}{n_+!(N-n_+)!} \right] \\
&= k \left[N \ln N - n_+ \ln n_+ - (N-n_+) \ln(N-n_+) \right],
\end{aligned}
\tag{4.28}
$$

where in the second step of this equation sequence we assume that $N \gg 1$ and $(N - n_+) = n_- \gg 1$. Given (4.26a),

$$
\begin{aligned}
\frac{S(E,N)}{k} &= N \ln N - \left(\frac{N}{2} - \frac{E}{2m_B B_o} \right) \ln \left(\frac{N}{2} - \frac{E}{2m_B B_o} \right) \\
&\quad - \left(\frac{N}{2} + \frac{E}{2m_B B_o} \right) \ln \left(\frac{N}{2} + \frac{E}{2m_B B_o} \right).
\end{aligned}
\tag{4.29}
$$

Note that, according to (4.29), $S(0,N) = kN \ln 2$.

Energy equation of state

Equation (4.29) tells us how the entropy S of the system depends upon its energy E. Thus, $1/T = (\partial S / \partial E)_{V,N}$ determines the energy equation of state

$$\frac{1}{kT} = \left(\frac{1}{2m_B B_o}\right)\left\{\ln\left(\frac{N}{2} - \frac{E}{2m_B B_o}\right) + 1 - \ln\left(\frac{N}{2} + \frac{E}{2m_B B_o}\right) - 1\right\}$$

$$= \left(\frac{1}{2m_B B_o}\right)\ln\left[\frac{1 - E/(Nm_B B_o)}{1 + E/(Nm_B B_o)}\right]. \tag{4.30}$$

Solving for $E/(Nm_B B_o)$ we find that

$$E = Nm_B B_o\left(\frac{1 - e^{2m_B B_o/kT}}{1 + e^{2m_B B_o/kT}}\right). \tag{4.31}$$

Note that when $B_o = 0$, the energy of the system $E = 0$. Also, as the temperature approaches zero, $T \to 0$, the energy approaches its minimum value $-Nm_B B_o$ with all paramagnet dipole moments parallel to the external field. Alternatively, very high temperatures, for which $2m_B B_o/kT \ll 1$, will return the system to zero energy, $E = 0$, and effectively demagnetize the system.

The mathematics of this model is, apart from an unimportant shift in the zero of energy, formally identical to that of a crystal with Shottky defects as described in Example 4.2. But the physics is significantly different. The energy of an empty lattice point, that is, of a Shottky defect, depends upon the extent to which the lattice points that surround the defect are filled. Therefore, the analysis of Example 4.2 applies only when the number of defects is small relative to the number of lattice points. Not so for the model of paramagnetism of (4.24)–(4.31). After all, the energy of each dipole depends only upon an intrinsic dipole moment and the strength of the applied field.

4.7 Negative absolute temperature

When the direction of the magnetic field applied to a paramagnet composed of spin-½ dipoles is suddenly reversed, an interesting situation develops. Consider a system of paramagnets with more parallel than antiparallel paramagnets so that $n_+ > n_-$ and, according to (4.25), $E < 0$. This situation arises naturally when a system of spin-½ paramagnetic dipoles is cooled. Now imagine that the direction of the applied field B_o is suddenly reversed. Then suddenly there are more antiparallel than parallel paramagnets so that $n_- > n_+$ and $E > 0$. This reversal is said to produce a *population inversion* of the magnetic dipoles.

A system of population-inverted magnetic dipoles is in a region of parameter space in which an increase in the system energy E further increases the already relatively large number n_- of antiparallel dipoles, and, therefore, decreases the system entropy S. Recall that when we know how the entropy S of a system

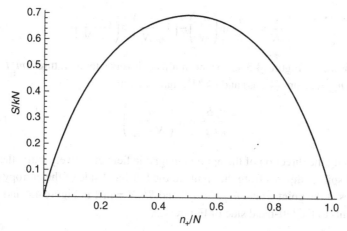

Figure 4.4 The normalized entropy S/kN of a system of spin-½ magnetic dipoles versus the fraction n_+/N of parallel, low energy, dipoles as described by Eq. (4.32).

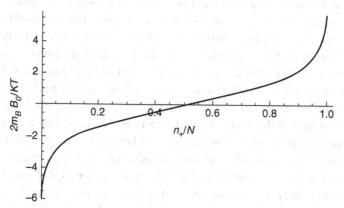

Figure 4.5 The normalized inverse temperature $2m_B B_o/kT$ of a system of spin-½ magnetic dipoles versus the fraction n_+/N of parallel dipoles as described by Eq. (4.33).

depends upon its energy E, we know the absolute temperature of that system through $1/T = (\partial S/\partial E)_{V,N}$. And because, in this case, $(\partial S/\partial E)_{V,N} < 0$, the absolute temperature of a population-inverted system of paramagnets is negative.

Figures 4.4 and Figure 4.5 illustrate these facts. Shown in Figure 4.4 is a plot of normalized entropy S/kN versus the proportion n_+/N of parallel dipoles from (4.28), that is, from

$$\frac{S}{kN} = -\frac{n_+}{N}\ln\left(\frac{n_+}{N}\right) - \left(1-\frac{n_+}{N}\right)\ln\left(1-\frac{n_+}{N}\right), \qquad (4.32)$$

and shown in Figure 4.5 is the normalized inverse temperature $2m_B B_o/kT$ versus n_+/N, from (4.26a) and (4.31), that is, from

$$\frac{2m_B B_o}{kT} = \ln\left(\frac{n_+}{N-n_+}\right). \qquad (4.33)$$

Reversing the direction of the applied magnetic field effectively shifts the system of spin-½ dipoles from the right- to the left-hand side of the entropy peak that lies at the point where $n_+/N = 1/2$ and $S/kN = \ln 2$ in Figure 4.4 and from the right- to the left-hand side of Figure 4.5.

Experimental realization

Edward Purcell (1912–1997) and his colleague Robert Pound (1919–2010) were the first to create a population inversion among magnetic dipoles. Their chosen system, the nuclear spins of lithium ions in a lithium fluoride crystal, behave qualitatively as illustrated in Figures 4.4 and 4.5. Initially their crystal was in thermal equilibrium with the laboratory environment and in equilibrium with an applied magnetic field of 0.01 tesla. The applied field was then reversed within a millisecond, and the lithium nuclei quickly came to equilibrium among themselves and with the newly oriented applied field. The nuclei remained in this unusual population-inverted equilibrium for a period of minutes before again equilibrating with the laboratory environment. Purcell and Pound demonstrated that this system had a negative temperature by shining monochromatic radiation with a frequency $v\ [= 2m_B B_o/h]$ matched to the transition between nuclear spin states. That the incident radiation was enhanced through stimulated emission of the nuclei rather than absorbed showed that the nuclear spins were preferentially antiparallel to the applied magnetic field – just as expected if the system of nuclear spins had a negative temperature. These beautiful experiments, first performed in 1950, have since become routine.

Negative temperatures can be achieved in any system composed of subsystems that (1) have a limited number of energy levels, (2) achieve equilibrium among themselves relatively quickly, and (3) equilibrate with their environment relatively slowly. For instance, the molecules that compose the active media in lasers and masers observe these three conditions. Such so-called "abnormal systems" seem exotic, but do not violate any law of thermodynamics. However, most systems are "normal" in the sense that their entropy S increases monotonically with their internal energy E and, therefore, have positive temperatures.

Problems

4.1 Oscillator entropy

The goal is to guide one to the efficient derivation of the entropy S of N one-dimensional, simple harmonic oscillators in terms of energy E and particle number N as found in (4.12). Start with (4.08) and (4.11) and (a) express hv_o/kT and Z_1 in terms of the dimensionless quantity $x = E/Nhv_o$. (b) Use these results and the relation $S(E,V) = c(N) + k[\beta E + N \ln Z_1]$ to find the entropy function $S(E,N)$ displayed in (4.12).

4.2 Nitrogen molecules

Diatomic molecules or dimers such as N_2 and O_2 can be modeled as simple harmonic oscillators with a characteristic frequency of oscillation v_o and energy levels $\varepsilon_n = hv_o(n + \frac{1}{2})$ where $n = 0,1,2,...$ A nitrogen dimer, for instance, is characterized by $hv_o = 0.3$ eV. What is the proportion of room-temperature (300 K) nitrogen molecules in their first $(n = 1)$ excited state? Note that k (300 K) = (1/40) eV.

4.3 Correspondence principle

Show that the Einstein solid, that is, a system of three-dimensional quantized, simple harmonic oscillators, observes the correspondence principle by showing (a) that its energy (4.10) reduces to the result $3NkT$ in the semi-classical limit and (b) that its entropy (4.14) reduces to $3Nk[1 + \ln(E/3Nhv_o)]$ in the semi-classical limit.

4.4 Two-level system

A system consists of N distinguishable, independent particles each of which can exist in only two states: one with energy 0 and the other with energy ε. Find (a) its single-particle partition function Z_1, (b) its energy E, and (c) its heat capacity $C [= dE/dT]$. (d) Plot or sketch the dimensionless quantities C/kN versus ε/kT from 0 to 3 and interpret the result.

4.5 Entropy maximum

(a) Show that, according to (4.32), the relative frequency of a system of spin-½ magnetic dipoles parallel to an applied magnetic field n_+/N that maximizes the entropy function $S(n_+)$ is given by $n_+/N = 1/2$. (b) Show that the maximum entropy of a system of spin-½ magnetic dipoles is given by $S = Nk\ln 2$.

4.6 Third law limit

Show that the entropy of a system of spin-½ paramagnets is third law compliant.

4.7 Chemical potential

An entropy that is a function $S(E,V,N)$ of three extensive variables E, V, and N implies the existence of three equations of state. We have made use of the energy $1/T = (\partial S/\partial E)_{V,N}$ and pressure $P/T = (\partial S/\partial V)_{E,N}$ equations of state but not yet the equation of state containing the *chemical potential* μ, that is, $\mu/T = -(\partial S/\partial N)_{E,V}$. The chemical potential μ is the energy added to the system per particle given that its volume and entropy are held constant, that is, $\mu = (\partial E/\partial N)_{S,V}$. Derive the chemical potential equation of state for a system described by the Sackur–Tetrode entropy (4.2).

4.8 Frequency of paramagnet orientations

A crystal lattice contains spin-½ paramagnets at each lattice site. (a) Derive expressions for the frequency of parallel paramagnet orientations n_+/N and antiparallel paramagnet orientations n_-/N in terms of the applied field B_o and temperature T. (Here and in Problem 4.9 you may find it convenient to employ the notation $x = 2m_B B_o/kT$.) (b) Using these expressions find the frequency of parallel and antiparallel orientations in the limits $T \to \infty$ and $T \to 0$. (c) Given that the value of a spin-½ magnetic moment is given by $m_B = eh/4\pi m_e$ ($= 0.927 \times 10^{-22}$ amp m^2) find the temperature at which 75% of the paramagnets are aligned with an applied field of $B_0 = 0.01$ tesla.

4.9 Heat capacity of a paramagnet

(a) Find the heat capacity $C\,[= dE/dT]$ of a paramagnet composed of N spin-½ dipoles with magnetic moment m_B in an externally applied magnetic field B_o. (b) What is the heat capacity C in the limits $T \to \infty$, $T \to -\infty$, and $T \to 0$?

4.10 Curie's law

The total magnetization M of a system of paramagnets is the sum of the individual magnetic dipole moments. When the paramagnets are spin-½ dipoles this sum is given by $M = (n_+ - n_-)m_B$. (a) Find an expression for M in terms of N, B_o, m_B, and T. (b) Show that the magnetization $M \propto T^{-1}$ when $2m_B B_o / kT \ll 1$. The relation $M = CB_o / T$, valid in this regime, is called *Curie's law*. (c) Find the constant C in Curie's law.

5

Entropy of a non-isolated system

5.1 Beyond the fundamental postulate

According to the fundamental postulate, *the microstates of an isolated system are equally probable*. While the fundamental postulate is quite powerful, it does not apply directly to systems that are not isolated. For instance, a very common non-isolated system is one in thermal equilibrium with its environment. When a system is not isolated we have no reason to believe that its microstates are equally probable. And when a system's microstates are not equally probable, the entropy will no longer be proportional to the logarithm of the multiplicity. In fact, we have considered such a system before: a single particle of gas in thermal equilibrium with the other particles in the gas at temperature T. According to the Maxwell–Boltzmann distribution the probability that this particle will occupy microstate i with energy ε_i is $e^{-\varepsilon_i/kT}/Z_1$ which is certainly not the probability that it will occupy a different microstate with a different energy. In this section we will describe the entropy of an arbitrarily complex system whose microstates are not equally probable and give a more general derivation of the Maxwell–Boltzmann distribution.

Ensembles

We proceed by constructing, in our imagination, a collection of N independent copies of the system. The members of this collection are identical except that n_j $(j = 1, 2, \ldots)$ of the copies in the collection occupy microstate j with a frequency,

$$\frac{n_j}{N} = p_j, \tag{5.1}$$

Figure 5.1 An ensemble of identical systems. Each system occupies a microstate indicated by the level on which the filled circle rests. Because the frequency with which a system occupies a microstate reflects a set of probabilities, the two middle microstates appear probabilistically favored.

that reflects the probability p_j that a copy will occupy its jth microstate. Of course, n_i and p_i are constrained so that

$$N = \sum_j n_j \tag{5.2}$$

and

$$1 = \sum_j p_j. \tag{5.3}$$

This collection of imaginary copies is called an *ensemble*. We owe the concept of an ensemble to J. Willard Gibbs (1839–1903). Its essential element is a set of microstate probabilities p_1, p_2, ...

5.2 The Gibbs entropy formula

Figure 5.1 illustrates a few members of an ensemble of identical systems in which each system occupies a microstate indicated by the level on which the filled circle rests. The systems shown in the figure appear to favor the middle two microstates. Since the members of an ensemble have been constructed in our imagination, we can tag or order them so that each is distinguishable from all others. For this reason we can apply the statistics of distinguishable objects to the members of an ensemble.

Now consider the N distinguishable systems that compose the ensemble as a single composite system. The entropy of a macrostate of this N-subsystem, composite system is a function $S(n_1, n_2, ...)$ of the occupation numbers $n_1, n_2, ...$ that define its macrostate. Since the systems that compose the ensemble are distinguishable, each of the $N!/(n_1! n_2! \cdots)$ distinct arrangements in which n_1 of the N subsystems occupy microstate 1, n_2 of the N subsystems occupy microstate 2, and so on is a distinct microstate of this occupation number macrostate of the composite system we call an ensemble.

And since by construction the ensemble itself is isolated, each of these distinct microstates is equally probable. Therefore, the entropy of this ensemble macrostate is given by

$$S(n_1, n_2, \cdots n_\Omega) - S(0, 0, \cdots N, \cdots 0) = k \ln \left[\frac{N!}{n_1! n_2! \cdots} \right]$$

$$= k \left\{ N \ln N - \sum_j n_j \ln n_j \right\}$$

$$= k \left\{ \sum_j n_j \ln N - \sum_j n_j \ln n_j \right\} \qquad (5.4)$$

$$= -k \sum_j n_j \ln \frac{n_j}{N}$$

$$= -kN \sum_j \frac{n_j}{N} \ln \frac{n_j}{N}$$

$$= -kN \sum_j p_j \ln p_j,$$

where we have used $n_j/N = p_j$ in the last step. The entropy $S(0, 0, \ldots N, \ldots, 0)$ is that of the ensemble in a macrostate consisting of a single microstate, that is, a macrostate in which each of the N subsystems occupies the same microstate. If we make the entropy $S(n_1, n_2, \ldots, n_\Omega)$ third law compliant, then

$$S(0, 0, \cdots, N, \cdots, 0) = 0. \qquad (5.5)$$

Furthermore, the additivity, more precisely the extensivity, of the entropy and $n_j = p_j N$ assure us that

$$S(n_1, n_2, \cdots, n_\Omega) = S(p_1 N, p_2 N, \cdots, p_\Omega N)$$

$$= NS(p_1, p_2, \cdots, p_\Omega) \qquad (5.6)$$

where $S(n_1, n_2, \cdots)$ is the entropy of the ensemble and $S(p_1, p_2, \cdots)$ is the entropy of one of its identical subsystems. These requirements (5.5) and (5.6) transform (5.4) into an expression for the absolute entropy

$$S(p_1, p_2, \cdots, p_\Omega) = -k \sum_{j=1}^{\Omega} p_j \ln p_j \qquad (5.7)$$

of a system that occupies its microstates with unequal probabilities $p_1, p_2, \ldots, p_\Omega$. This result, sometimes called the *Gibbs entropy formula*, reduces to $S = k \ln \Omega$ when $p_1 = p_2 = \cdots = p_\Omega = 1/\Omega$.

Random variables

Note that, according to $S = -k\sum_j p_j \ln p_j$, the dynamical variables of a non-isolated system must be *random variables* because the system realizes its microstates according to a set of probabilities p_j. In particular, the energy of a non-isolated system is a random variable because it realizes a range of values $E_1, E_2, \ldots$ according to a set of probabilities $p_1, p_2, \ldots$. Up until now we have considered only isolated systems that have a definite energy. The transition from systems described with so-called *sure variables* to systems described with random variables is significant.

A random variable has an average value that is commonly called an *expectation value*. For instance, the expectation value of a non-isolated system's energy, denoted $\langle E \rangle$, when that system realizes its microstate energy E_j with probability p_j is defined by

$$\langle E \rangle = \sum_j p_j E_j. \tag{5.8}$$

Likewise, the pressure is also a random variable that assumes values P_j $\left[= -\partial E_j / \partial V \right]$ with probabilities p_j. Thus,

$$\langle P \rangle = \sum_j p_j P_j = -\sum_j p_j \left(\partial E_j / \partial V \right). \tag{5.9}$$

Expectation values are the observables of a thermodynamic system. When the system is isolated the expectation values $\langle E \rangle$ and $\langle P \rangle$ reduce to the sure variables E and P. Texts that make extensive use of probabilistic mechanics usually forgo special notation for expectation values and simply denote $\langle E \rangle$ and $\langle P \rangle$ with E and P – a possibly confusing abbreviation. The entropy $S(p_1, p_2, \ldots, p_\Omega)$ $\left[= -k\sum_j p_j \ln p_j \right]$ is itself neither a random variable nor an expectation value but simply a function (5.7) of microstate probabilities.

Eponymy

Finally, we observe that the names *Boltzmann entropy* $S \left[= k \ln \Omega \right]$ and *Gibbs entropy formula* $S = -k\sum_j p_j \ln p_j$ are historically misleading. The history is complex but it is probably fair to say that Boltzmann discovered a continuous version of the Gibbs entropy formula while Planck introduced the Boltzmann entropy. This history reminds one of *Stigler's law of eponymy*: *No discovery is named after its original discoverer*.

5.3 Canonical ensemble

A canonical ensemble is one defined by probabilities $p_1, p_2, \ldots$ that characterize the microstates of a system in equilibrium with an environment at temperature T. These probabilities maximize the Gibbs entropy $S \left[= -k \sum_j p_j \ln p_j \right]$ they determine given the normalization (5.3) and an expectation value $\langle E \rangle$ $\left[= \sum_j E_j p_j \right]$. Maximizing the constrained entropy

$$-\sum_j p_j \ln p_j + \alpha \left[1 - \sum_j p_j \right] + \beta \left[\langle E \rangle - \sum_j p_j E_j \right] \qquad (5.10)$$

with respect to p_i produces

$$- \ln p_i - 1 - \alpha - \beta E_i = 0, \qquad (5.11)$$

whose solution is the probability distribution

$$p_i = e^{-(1+\alpha)} e^{-\beta E_i}. \qquad (5.12)$$

Since the second derivatives of (5.10) with respect to the microstate probabilities p_i are negative, (5.12) identifies a relative maximum and not merely a stationary value of the constrained missing information.

The Lagrange multipliers α and β are, in general, determined by the probability distribution (5.12) and the two constraints (5.3) and (5.8). Requiring that the probability distribution (5.12) satisfy the normalization condition (5.3) yields

$$e^{-(1+\alpha)} = \frac{1}{\sum_i e^{-\beta E_i}} \qquad (5.13)$$

$$= \frac{1}{Z},$$

where the system partition function

$$Z = \sum_i e^{-\beta E_i} \qquad (5.14)$$

is a sum over the microstates of the system. Thus, the system partition function Z generalizes the single-particle partition function Z_1 used in Chapters 3 and 4. Given (5.13) the probability distribution (5.12) becomes

$$p_i = \frac{e^{-\beta E_i}}{\sum_j e^{-\beta E_j}}$$

$$= \frac{e^{-\beta E_i}}{Z},$$

(5.15)

and the energy constraint (5.8) is equivalent to

$$\langle E \rangle = -\frac{\partial \ln Z}{\partial \beta}.$$

(5.16)

Together the probability distribution (5.15) and the Gibbs entropy formula $S = -\sum_j p_j \ln p_j$ imply that

$$S = k\left[\beta\langle E \rangle + \ln Z\right].$$

(5.17)

Consequently, the derivative $\left(\partial S / \partial \langle E \rangle\right)\left[= 1/T\right]$ produces a result

$$\frac{1}{kT} = \beta + \left[\langle E \rangle + \frac{\partial \ln Z}{\partial \beta}\right]\left(\frac{\partial \beta}{\partial \langle E \rangle}\right)$$

$$= \beta$$

(5.18)

that turns the probability distribution (5.15) into the so-called *Boltzmann distribution*

$$p_i = \frac{e^{-E_i/kT}}{Z},$$

(5.19)

where $Z = \sum_j e^{-\beta E_j}$ and $\beta = 1/kT$.

This derivation of the Boltzmann distribution (5.19) is formally identical to the derivation of the single-particle Maxwell–Boltzmann distribution in Section 3.4, but the context is much broader. The microstate energies E_i in (5.19) are those of any system however complex or simple and the sum in the partition function (5.14) is over all its microstates. If one first determines the partition function $Z = \sum_j e^{-\beta E_j}$, then the energy equation of state $\langle E \rangle = -\partial \ln Z/\partial \beta$, the entropy $S = k\left[\beta\langle E \rangle + \ln Z\right]$, and the Boltzmann distribution $p_j = e^{-\beta E_j}/Z$, where $\beta = 1/kT$, all follow. The partition function is a flexible and straightforward approach to statistical mechanics that avoids multiplicities and combinatoric formulas.

5.4 Partition functions

The systems we consider in this book are composed of identical particles that occupy their single-particle microstates independently apart from system-wide constraints. In this case the energy E_h of a particular system in microstate h is a simple sum,

$$E_h = \varepsilon_{1,i} + \varepsilon_{2,j} + \cdots + \varepsilon_{N,k}, \tag{5.22}$$

of the energies, $\varepsilon_{1,i}, \varepsilon_{2,j}, \ldots \varepsilon_{N,k}$ of the N particles that form the system when in this microstate. Here, for instance, the subscript $2, k$ would indicate the energy $\varepsilon_{2,k}$ of the second particle when in its kth single-particle microstate. The subscripts $i, j, \ldots, k$ together define the system microstate h. Therefore, the system partition function is given by

$$
\begin{aligned}
Z &= \sum_h e^{-\beta E_h} \\
&= \sum_h e^{-\beta \varepsilon_{1,i} - \beta \varepsilon_{2,j} \cdots - \beta \varepsilon_{N,k}} \\
&= \sum_{i,j,\cdots,k} e^{-\beta \varepsilon_{1,i}} e^{-\beta \varepsilon_{2,j}} \cdots e^{-\beta \varepsilon_{N,k}},
\end{aligned} \tag{5.23}
$$

where the sum is over all values of the N single-particle indices $i, j, \ldots, k$ that determine the system microstate.

When in addition to their independence the identical particles that compose the system are distinguishable from one another, each realization of one particle's single-particle microstate combines with each realization of all the other particles' single-particle microstates to make a distinct N-particle system microstate. Then the partition function sum (5.23) is further decomposed:

$$
\begin{aligned}
Z &= \sum_{i,j,\cdots k} e^{-\beta \varepsilon_{1,i}} e^{-\beta \varepsilon_{2,j}} \cdots e^{-\beta \varepsilon_{N,k}} \\
&= \sum_i \sum_j \cdots \sum_k e^{-\beta \varepsilon_i} e^{-\beta \varepsilon_j} \cdots e^{-\beta \varepsilon_k} \\
&= \sum_i e^{-\beta \varepsilon_j} \sum_j e^{-\beta \varepsilon_k} \cdots \sum_i e^{-\beta \varepsilon_k} \\
&= \left(\sum_i e^{-\beta \varepsilon_i} \right)^N \\
&= Z_1^N.
\end{aligned} \tag{5.24}
$$

Given this result, those equations that encapsulate the partition function method applied to a complex system (5.14), (5.16), (5.17), and (5.19) reproduce similar equations in terms of the single-particle partition function as derived in Section 3.4.

Equations (5.14), (5.16), (5.17), and (5.19) also reduce to familiar results in the special case when all the microstates of the system are degenerate, that is, when all microstates have the same energy E. In this case

$$Z = \sum_j e^{-\beta E_i}$$
$$= \Omega e^{-\beta E},$$

(5.25)

where Ω is the number of system microstates with E. The resulting entropy,

$$S = k\left[\beta E + \ln Z\right]$$
$$= k\left[\beta E + \ln \Omega - \beta E\right]$$
$$= k \ln \Omega,$$

(5.26)

is, as expected, that of an isolated system, or alternatively, that characteristic of a system in a so-called *microcanonical ensemble*.

We will not, in this Guide, advance the random variable view of statistical mechanics much further. But see Problems 5.1 and 5.2 and Sections 8.4 and 8.7.

5.5 Entropy metaphors

The Boltzmann (5.26) and Gibbs entropy (5.7) formulas provide us with formal answers to the question "What is entropy?" But they also suggest appropriate metaphors that summarize the content of the entropy concept. For instance, the Boltzmann entropy $S\left[= k \ln \Omega\right]$ suggests that entropy is *spread in phase space*. For an isolated system whose macrostate consists of many microstates, that is, one for which $\Omega \gg 1$, probabilistically inhabits a correspondingly large volume of phase space, and an isolated system that occupies only one microstate, so that $\Omega = 1$, is as confined in phase space as can be.

According to the Gibbs formula the entropy is a function of the probabilities with which a system occupies its accessible microstates. This function is largest when the microstate probabilities are equiprobable, that is, when $p_1 = p_2 = \cdots$ and smallest when one microstate is certain, say, when $p_1 = 1$ and $p_2 = p_3 = \cdots = 0$. Thus, an event whose possible outcomes are equally probable is maximally uncertain and one whose possible outcomes have vanishing probability except for one with probability one is minimally uncertain. In this sense, the Gibbs entropy $S\left[=-k\sum_j p_j \ln p_j\right]$ is a measure of the uncertainty of a system's microstate. Thus, the metaphor *uncertainty* also conveys the meaning of entropy.

Problems

5.1 Canonical ensemble of Shottky defects

The context is Example 4.2. The system is a lattice site within a crystalline array of lattice sites in thermal equilibrium with each other at temperature T. When occupied by an atom or molecule the lattice site has energy $E_0 = 0$ and when empty the lattice site has energy $E_1 = \varepsilon$. (a) What is the probability p_0 that the lattice site will be occupied and the probability p_1 that the lattice site will be unoccupied? (b) Use these probabilities to determine the entropy S_1 $\left[= -k\sum_i p_i \ln p_i \right]$ of a single lattice site. (c) Show that the entropy of N lattice sites $S \left[= NS_1 \right]$ is consistent, with the entropy S of a crystal with N sites containing n Shottky defects as determined in Example 4.2.

5.2 The Gibbs entropy formula

Consider a system in equilibrium with its environment at temperature T. The probability that this system occupies microstate i is given by $p_i = e^{-\beta E_i}/Z$ where E_i is the energy of the system when in microstate i and its partition function $Z = \sum_j e^{-\beta E_j}$. From this result and the Gibbs entropy formula derive Eq. (5.17) for the entropy S of the system and Eq. (5.16) for the expected energy $\langle E \rangle$ $\left[= \sum_i p_i E_i \right]$ of the system expressed in terms of the system partition function Z and the factor $\beta \left[= 1/kT \right]$.

6

Entropy of fermion systems

6.1 Symmetries and wave functions

An important difference between the classical and quantum perspectives is their different criteria of distinguishability. Identical particles are classically distinguishable when separated in phase space. On the other hand, identical particles are always quantum mechanically indistinguishable for the purpose of counting distinct microstates. But these concepts and these distinctions do not tell the whole story of how we count the microstates and determine the multiplicity of a quantized system.

There are actually two different ways of counting the accessible microstates of a quantized system of identical, and so indistinguishable, particles. While these two ways were discovered in the years 1924–1926 independently of Erwin Schrödinger's (1887–1961) invention of wave mechanics in 1926, their most convincing explanation is in terms of particle wave functions. The following two paragraphs may be helpful to those familiar with the basic features of wave mechanics.

A system of identical particles has, as one might expect, a probability density that is symmetric under particle exchange, that is, the probability density is invariant under the exchange of two identical particles. But here wave mechanics surprises the classical physicist. A system wave function may either keep the same sign or change signs under particle exchange. In particular, a system wave function may be either symmetric or antisymmetric under particle exchange.

For this reason identical particles come in two and only two kinds: *bosons* (for instance, phonons, photons, and helium-4 atoms) described by symmetric wave functions and *fermions* (for instance, electrons, protons, neutrons, and helium-3 atoms) described by antisymmetric wave functions. A consequence

105

of the different symmetry properties of bosons and fermions and of the fact that wave functions superpose is that the probability density of a system of bosons is non-zero where two bosons inhabit the same single-particle microstate and the probability density of a system of fermions must vanish where two fermions are imagined to occupy the same single-particle microstate. Thus there are two kinds of identical particles: those that can occupy the same single-particle microstate (bosons) and those that cannot (fermions).

The spin-statistics theorem and the Pauli principle

The *spin-statistics theorem*, discovered by Wolfgang Pauli (1900–1958) in 1940, builds upon these features of wave mechanics. According to the spin-statistics theorem, all particles with even units of half-integer spin are bosons and all particles with odd units of half integer spin are fermions. Spin is intrinsic angular momentum. The unit of spin is the reduced Planck constant $\hbar\left[= h/2\pi\right]$. Therefore, one unit of half-integer spin is $\hbar/2$. Pauli had in 1924 proposed the rule that electrons within an atom cannot occupy the same single-particle microstate – the so-called *Pauli principle* – without at the time realizing that this behavior was characteristic of a larger class of particles we now call fermions.

We summarize. The only important distinction for the statistical mechanics of quantized systems is the distinction between bosons, which can share the same single-particle microstate, and fermions, which cannot. Bosons are said to obey Bose–Einstein statistics, named after the two men, Satyendra Nath Bose (1894–1974) and Albert Einstein, who in 1924 independently discovered (Bose) and applied (Bose and Einstein) this way of counting distinct microstates. Fermions obey Fermi–Dirac statistics named after the two men, Enrico Fermi (1901–1954) and P. A. M. Dirac (1902–1984), who in 1926 independently discovered and applied this way of counting distinct microstates. In the previous chapter we investigated systems of bosons (phonons) and continue to do so in the next (photons and bosons with non-zero rest mass). Here we explore the statistical mechanics of fermions.

6.2 Intrinsic semiconductors

The single-particle microstates available to the electrons of a material in its condensed phase are of two kinds: bound or valence band states and conduction band states. Since electrons are fermions, at most one electron at a time

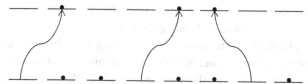

Figure 6.1 Eight valence band microstates and eight conduction band single-particle microstates. Three electrons have been excited from the valence band to the conduction band.

can occupy a bound or conduction band single-particle microstate. Typically the energies of the bound microstates crowd together, as do the energies of the conduction band microstates, while a *bandgap* of up to several electronvolts may separate these two groups or *bands* of single-particle microstate energies. The exact size of the bandgap determines the conductivity of a material. The bandgap is relatively large for insulators, relatively small for semiconductors, and nonexistent for conductors.

The most common solid-phase semiconducting material is crystalline silicon. An *intrinsic semiconductor* is a crystal with no defects or impurities. When lattice vibrations excite a valence electron into the conduction band of an intrinsic semiconductor, both the vacancy or *hole* left in the valence band and the newly formed conduction electron are free to move under the influence of an applied electric field. Holes and electrons contribute in equal measure to the conductivity of an intrinsic semiconductor.

Figure 6.1 illustrates the simplified model we consider. The numbers of valence and conduction band single-particle microstates are equal as are the number of holes and conduction electrons. In this illustration the valence and conduction band single-particle microstates have been made degenerate with one energy level for each band. The valence and conducting bands of a more complete model would consist of closely spaced non-degenerate energy levels. This model captures one important feature of intrinsic semiconductors – bandgap energy – at the cost of ignoring other features.

Entropy

The arrangement of holes in the valence band and the arrangement of electrons in the conduction band both contribute to the multiplicity of an intrinsic semiconductor. Suppose the semiconductor has N electrons, N valence band single-particle microstates, and N conduction band single-particle microstates. Of these N electrons, n occupy the conduction band and, because the number of electrons is conserved, $N - n$ are left to occupy the valence band. Also suppose the two bands are separated by bandgap energy ε.

Because electrons are fermions, no more than one electron can occupy each single-particle microstate. When n electrons are excited from the valence band to the conduction band, the N valence band, single-particle microstates divide into two groups: one group of n microstates that are unoccupied holes and another group of $N - n$ microstates that are occupied by one electron. Therefore, the binomial coefficient $N!/[n!(N-n)!]$ gives the number of, equally probable, arrangements of the valence band subsystem. Similarly, the N conduction band microstates divide into two exclusive groups: n are occupied by a single electron and $N - n$ are unoccupied. Therefore, the number of equally probable arrangements of the conduction band subsystem is also $N!/[n!(N-n)!]$. Because each valence band subsystem arrangement can be paired with each conduction band subsystem arrangement to form a single system microstate, the multiplicity of the intrinsic semiconductor system is

$$\Omega = \left[\frac{N!}{n!(N-n)!}\right]^2 \tag{6.1}$$

and therefore its entropy is given by

$$
\begin{aligned}
S(n,N) &= k \ln\left[\frac{N!}{n!(N-n)!}\right]^2 \\
&= 2k\left[N \ln N - n \ln n - (N-n)\ln(N-n)\right] \\
&= 2kN\left[-\left(\frac{n}{N}\right)\ln\left(\frac{n}{N}\right) - \left(1-\frac{n}{N}\right)\ln\left(1-\frac{n}{N}\right)\right],
\end{aligned}
\tag{6.2}
$$

where we have assumed that $N - n \gg 1$ and $n \gg 1$.

Since the energy of the system, apart from an inconsequential constant, is

$$E = n\varepsilon, \tag{6.3}$$

where ε is the bandgap energy and, of course, n is the number of conduction band electrons, the entropy S of the intrinsic semiconductor as a function of system energy E is given by

$$S(E,N) = 2kN\left[-\left(\frac{E}{N\varepsilon}\right)\ln\left(\frac{E}{N\varepsilon}\right) - \left(1-\frac{E}{N\varepsilon}\right)\ln\left(1-\frac{E}{N\varepsilon}\right)\right]. \tag{6.4}$$

The energy equation of state

We may now determine the energy equation of state from (6.4) and $1/T = (\partial S/\partial E)_N$. In this way, we find that

$$\frac{1}{kT} = 2N\left[-\left(\frac{1}{N\varepsilon}\right)\ln\left(\frac{E}{N\varepsilon}\right) - \left(\frac{1}{N\varepsilon}\right) + \left(\frac{1}{N\varepsilon}\right)\ln\left(1 - \frac{E}{N\varepsilon}\right) + \left(\frac{1}{N\varepsilon}\right)\right]$$

$$= \frac{2}{\varepsilon}\left[-\ln\left(\frac{E}{N\varepsilon}\right) + \ln\left(1 - \frac{E}{N\varepsilon}\right)\right] \tag{6.5}$$

$$= \frac{2}{\varepsilon}\ln\left(\frac{N\varepsilon}{E} - 1\right).$$

Solving (6.5) for the energy E produces the energy equation of state

$$E = \frac{N\varepsilon}{1 + e^{\varepsilon/2kT}}. \tag{6.6}$$

That the frequency of occupied conduction band, single-particle microstates is given by

$$\frac{n}{N} = \frac{1}{1 + e^{\varepsilon/2kT}} \tag{6.7}$$

follows immediately from (6.3) and (6.6). As one might expect, in the limit of vanishingly small absolute temperature the frequency of conduction band electrons vanishes, that is, $n/N \to 0$ as $\varepsilon/2kT \to \infty$. In the limit of indefinitely large absolute temperature the electrons are shared equally between valence and conduction bands, that is, $n/N \to 1/2$ as $\varepsilon/2kT \to 0$, also as one might expect.

Typical semiconductor bandgap energies are on the order of a few electron-volts ($1\,eV = 1.60 \times 10^{-19}$ joules). When, for instance, the bandgap energy $\varepsilon = 2\,eV$, the frequency of occupied conduction band single-particle microstates n/N, at room temperature ($T = 300$ K and $kT = 0.026$ eV), is given by

$$\frac{n}{N} = \frac{1}{1 + e^{1/0.026}}$$

$$= \frac{1}{1 + e^{38.5}} \tag{6.8}$$

$$= 1.9 \times 10^{-17}.$$

Thus, typically, an incredibly small fraction of the electrons and their holes contribute to the conductivity of an intrinsic semiconductor at room temperature.

Because the bandgap energy ε and the absolute temperature T enter into the argument of the exponent in expression (6.7), the frequency of occupied conduction band single-particle microstates n/N is quite sensitive to small changes in ε and T.

6.3 Ideal Fermi gas

We have used two different methods to derive the entropy of an ideal gas and the equations of state that follow from that entropy in Chapter 3: (1) the direct method in which one counts the number of microstates accessible to a gas of N particles that share energy E and occupy volume V, and (2) the occupation number method in which one maximizes the number of ways of distributing particles among the cells that compose a single-particle phase space while conserving N and E. These methods as applied in Chapter 3 share at least two distinctly classical assumptions: (1) identical particles are considered distinguishable and (2) the entropy is given by $S = c + k \ln \Omega$ where the "constant" c is chosen to make the entropy $S(E,V,N)$ an extensive function of its variables. These methods produce an extensive entropy and the familiar classical, ideal gas equations of state.

We are now in a position to revisit the ideal gas from the quantum perspective. We implement this perspective in four ways: (1) we quantize phase space with Planck's constant h; (2) we drop the classical assumption that identical particles are distinguishable; (3) we use the third law compliant absolute entropy $S = k \ln \Omega$; and (4) we assume in this chapter that the identical particles are fermions and in the next that they are bosons. Given these assumptions we will find that the entropies of the ideal Fermi and ideal Bose gases are extensive functions $S(E,V,N)$ that in one limit reproduce the Sackur–Tetrode entropy of an ideal classical gas while more generally retaining quantum features.

Because we retain the idealization that there are no interparticle forces, we continue to speak of an ideal gas: in this chapter of an ideal Fermi gas and in the next of an ideal Bose gas. We also simplify our investigation by assuming that each particle of the gas possesses the average energy E/N of the gas system. This *average energy approximation* produces results that, while approximate, describe important features of the ideal Fermi and ideal Bose gases. Other approaches to the ideal quantum gases that incorporate more physics can be found in the books listed in the Annotated Further Reading in Appendix VI.

The multiplicity of distinguishable particles

First a review. Recall from equation (3.15) that the multiplicity Ω of an ideal gas composed of N distinguishable particles that occupy volume V and share energy E is given by

$$\Omega(E,V,N) = \left[V \left(\frac{E}{N} \right)^{3/2} \left(\frac{4\pi em}{3h^2} \right)^{3/2} \right]^N . \tag{6.9}$$

Recall also that the symbol e stands, in this context, for the base of the natural logarithm. The factor in the brackets on the right hand side of (6.9) can be expressed as

$$\Omega(E/N,V,1) = V \left(\frac{E}{N} \right)^{3/2} \left(\frac{4\pi em}{3h^2} \right)^{3/2} , \tag{6.10}$$

in which case (6.9) becomes

$$\Omega(E,V,N) = \left[\Omega(E/N,V,1) \right]^N . \tag{6.11}$$

If $\Omega(E,V,N)$ is the multiplicity of an ideal gas composed of N distinguishable particles, possessing energy E, and occupying volume V, then $\Omega(E/N,V,1)$ must be the multiplicity of a single particle of ideal gas, possessing energy E/N, and occupying volume V when that particle is part of a larger system of N distinguishable particles. According to this interpretation a single particle of ideal gas may occupy $\Omega(E/N,V,1)$ equally probable microstates, a system of two distinguishable ideal gas particles may occupy $\left[\Omega(E/N,V,1) \right]^2$ equally probable microstates, and so on until a system of N distinguishable particles of ideal gas may occupy $\left[\Omega(E/N,V,1) \right]^N$ equally probable microstates.

The multiplicity of an ideal Fermi gas

The multiplicity $\Omega(E/N,V,1)$ of a single particle of ideal gas with energy E/N is, of course, the number of cells that can be occupied by that particle. Furthermore, *the number of cells that can be occupied by a single particle of ideal gas must be independent of whether that particle is itself distinguishable or indistinguishable from the other particles that compose the gas system and if indistinguishable whether fermion or boson.* This crucial insight allows us to proceed in the following way.

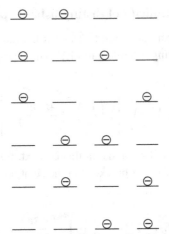

Figure 6.2 The six distinct ways of filling four cells with two fermions.

First, we establish the following notation. Let n stand for the number of cells that can be occupied by a particle of ideal gas with energy E/N, that is, let

$$n = \Omega(E/N, V, 1)$$
$$= V\left(\frac{E}{N}\right)^{3/2}\left(\frac{4\pi em}{3h^2}\right)^{3/2} \tag{6.13}$$

where in (6.13) we have used (6.10). Second, we count the number of distinct ways of filling n cells with $N\,[\leq n]$ fermions while leaving $n-N$ of them empty.

The problem of filling n cells with $N\,[\leq n]$ identical fermions has already been solved and exploited in Section 6.2 even if there the symbols n and N had different meanings. As an example consider the number of distinct ways of filling four cells with two fermions as illustrated in Figure 6.2. There are $4!/\left[(4-2)!2!\right]\,[=6]$ distinct ways. In general there are "n choose N" or

$$\Omega = \frac{n!}{(n-N)!N!} \tag{6.14}$$

distinct ways of filling n cells with N fermions.

The multiplicity described by (6.13) and (6.14) is the result of the average energy approximation. According to this approximation each gas particle possesses the same energy: the average energy E/N of the particles in the system. That the average energy approximation produces the exact multiplicity n^N when identical particles are considered distinguishable encourages us to use this approximation for the quantum ideal gases when identical particles are indistinguishable.

The entropy

Given (6.14) the entropy of an ideal Fermi gas is

$$\frac{S}{Nk} = \left(\frac{1}{N}\right)\ln\left[\frac{n!}{(n-N)!N!}\right]$$

$$= \frac{n}{N}\ln n - \left(\frac{n}{N}-1\right)\ln(n-N) - \ln N \qquad (6.15)$$

$$= -\left(\frac{n}{N}\right)\ln\left[1-\frac{N}{n}\right] + \ln\left[\frac{n}{N}-1\right]$$

where in the first step we have assumed that $N \gg 1$ and $n-N \gg 1$ in order to use Stirling's approximation. Also, recall that the number of single-particle cells is given by (6.13) and that the number of fermions N can be no larger than the number of cells n. Thus,

$$0 < \frac{N}{n} \leq 1 \qquad (6.16)$$

describes the domain of the average occupation number per cell N/n or, more briefly, the *occupancy* of a system of identical fermions.

This domain (6.16) reminds us that the ideal Fermi gas may not always realize one of the conditions, $n-N \gg 1$, assumed in deriving the entropy (6.15). In fact, at very low temperatures the occupancy N/n of an ideal Fermi gas approaches 1. For this reason, the assumption that $n-N \gg 1$ limits the model that follows from the entropy (6.15). In actual fact, the approximation $n-N \gg 1$ does little harm – probably because the form of Stirling's approximation we use, $\ln(n-N)! \approx (n-N)\ln(n-N) - (n-N)$, is observed not only when $(n-N) \gg 1$ but also in the limit as $n-N \to 0$.

Note that the number of cells n given by (6.13) is itself an extensive function of the fluid variables E, V, and N. And since

$$\frac{N}{n} = \left(\frac{N}{V}\right)\left(\frac{N}{E}\right)^{3/2}\left(\frac{3h^2}{4\pi em}\right)^{3/2} \qquad (6.17)$$

is the ratio of two extensive variables, the occupancy N/n is an intensive variable. When (6.17) is used to replace N/n in the normalized entropy S/Nk (6.15), we produce an extensive entropy function $S(E,V,N)$. This function describes all there is to know about the thermodynamic behavior of an ideal Fermi gas. However, we choose to keep the occupancy N/n as an auxiliary parameter because doing so simplifies the algebra. While the entropy (6.15) and the occupancy (6.17) solve the formal problem of deriving the entropy of the ideal Fermi gas in terms of its extensive variables, the particular properties of this gas remain to be investigated.

Low occupancy regime

When the average occupancy N/n is low, that is, when

$$\frac{N}{n} \ll 1, \tag{6.18}$$

the entropy (6.15) becomes

$$
\begin{aligned}
\frac{S(E,V,N)}{kN} &= 1 + \ln\left[\frac{n}{N}\right] \\
&= 1 + \ln\left[\left(\frac{V}{N}\right)\left(\frac{E}{N}\right)^{3/2}\left(\frac{4me\pi}{3h^2}\right)^{3/2}\right] \\
&= \frac{5}{2} + \ln\left[\left(\frac{V}{N}\right)\left(\frac{E}{N}\right)^{3/2}\left(\frac{4\pi m}{3h^2}\right)^{3/2}\right],
\end{aligned}
\tag{6.19}
$$

where in the first step of this equation sequence we have used the occupancy (6.17). This result (6.19) is the Sackur–Tetrode entropy (4.2) of an ideal gas. Occupancies for which $N/n \ll 1$ define the semi-classical approximation of an ideal Fermi gas.

Equations of state

We repeat: the entropy (6.15) and the occupancy (6.17) completely describe the thermodynamics of an ideal Fermi gas. In particular, its equations of state follow from the entropy (6.15), the occupancy (6.17), $1/T = (\partial S/\partial E)_{V,N}$, and $P/T = (\partial S/\partial V)_{E,N}$. From these we find that

$$\frac{E}{NkT} = -\left(\frac{3}{2}\right)\left(\frac{n}{N}\right)\ln\left(1 - \frac{N}{n}\right) \tag{6.20}$$

and

$$\frac{PV}{NkT} = -\left(\frac{n}{N}\right)\ln\left(1 - \frac{N}{n}\right). \tag{6.21}$$

An immediate consequence of (6.20) and (6.21) is that

$$E = \frac{3}{2}PV. \tag{6.22}$$

Note that, given the occupancy (6.17), Planck's constant h survives in the equations of state (6.20) and (6.21) and that assuming low occupancy $[N/n \ll 1]$ reduces them to the familiar $E = 3NkT/2$ and $PV = NkT$.

Low temperature regime

How does the ideal Fermi gas behave when the occupancy N/n is relatively high, especially when the occupancy N/n approaches its maximum value of 1? We are particularly concerned to know if the relatively high occupancy regime is equivalent to the low temperature regime and, if so, whether the ideal Fermi gas observes the third law.

While we cannot solve (6.15), (6.17), and (6.20) analytically for the function $S(T)$, we can use these equations to parameterize S and T in terms of the occupancy N/n for a given particle number N and volume V. Then we can use these parametric equations to plot $S(T)$.

In order to make our plot as meaningful as possible, we search for a normalization of T that casts Eqs. (6.17) and (6.20) into simplest form. In doing so we use E_0 to denote the energy realized when every one of the single particle cells is filled with one fermion, that is when $N = n$, and the volume remains at V. According to (6.13) or (6.17) E_0 is the lowest possible internal energy: the ground state energy. Thus, according to (6.13), we have

$1 = (V/N)(E_0/N)^{3/2}(4\pi em/3h^2)^{3/2}$ or, equivalently,

$$E_0 = \frac{N^{5/3}}{V^{2/3}}\left(\frac{3h^2}{4\pi em}\right). \tag{6.23}$$

Then (6.17) may be rewritten as

$$\frac{E}{E_0} = \left(\frac{n}{N}\right)^{2/3}. \tag{6.24}$$

Using (6.24) to eliminate energy E from the energy equation of state (6.20) produces

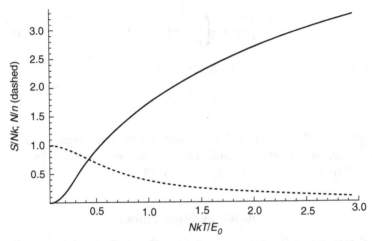

Figure 6.3 Normalized entropy S/Nk (solid line) and occupancy N/n (dashed line) versus normalized temperature NkT/E_0 when N and V are constant for the ideal Fermi gas as described by (6.15), (6.17), and (6.25).

$$\frac{NkT}{E_0} = -\left(\frac{2}{3}\right)\frac{(N/n)^{1/3}}{\ln(1-N/n)}. \tag{6.25}$$

This result and expression (6.15) for the normalized entropy S/Nk allow us to generate values of S/Nk and N/n as a function of NkT/E_0 and to plot the result. As we see in Figure 6.3, because $S/Nk \rightarrow 0$ as $T \rightarrow 0$ for constant values of V and N, the ideal Fermi gas observes the third law.

Heat capacity

The heat capacity at constant volume C_V of a fluid that obeys the fundamental constraint $dE = T\,dS - P\,dV$, is related to the entropy S by

$$C_V = T\left(\frac{\partial S}{\partial T}\right)_V. \tag{6.26}$$

And since the entropy of an ideal Fermi gas is a function of its temperature through (6.15), and (6.25), we find, after some algebra, that

$$\frac{C_V}{Nk} = \frac{3(n/N)\ln(1-N/n)}{\left[1+3(N/n)/\{(1-N/n)\ln(1-N/n)\}\right]}. \tag{6.27}$$

Furthermore, given (6.22) and (6.24), the normalized pressure PV/E_0 is given by

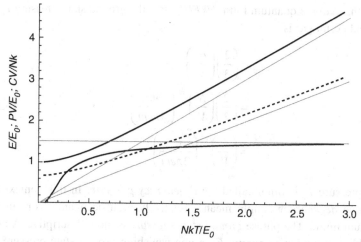

Figure 6.4 The normalized energy E/E_0 (upper solid), normalized pressure PV/E_0 (dashed), and normalized heat capacity C_V/Nk (lower solid) of an ideal Fermi gas from, respectively, (6.24), (6.28), and (6.27), given the relation (6.25) between normalized temperature NkT/E_0 and occupancy N/n. The thin lines represent classical limits. The volume V and particle number N are constant.

$$\frac{PV}{E_0} = \left(\frac{2}{3}\right)\left(\frac{n}{N}\right)^{2/3}. \tag{6.28}$$

Figure 6.4 summarizes the behavior of the normalized energy E/E_0 (6.24), the normalized pressure PV/E_0 (6.28), and the normalized heat capacity C_V/Nk (6.27) of an ideal Fermi gas given the relation (6.25) that parameterizes the normalized temperature NkT/E_0 in terms of the occupancy N/n.

Ground state

As the normalized temperature NkT/E_0 passes from values high compared to 1 to values low compared to 1, the ideal Fermi gas passes from its classical regime to its quantum regime. Because, given (6.23), the normalized temperature is

$$\frac{NkT}{E_0} = kT\left(\frac{V}{N}\right)^{2/3}\left(\frac{4\pi em}{3h^2}\right), \tag{6.29}$$

low temperature, high particle density, and small particle mass all contribute to placing the ideal Fermi gas in the quantum regime.

In the extreme quantum limit $NkT/E_0 \ll 1$ the ground state pressure P_0 of an ideal Fermi gas is

$$
\begin{aligned}
P_0 &= \left(\frac{2}{3}\right)\left(\frac{E_0}{V}\right) \\
&= \left(\frac{2}{3}\right)\left(\frac{N}{V}\right)^{5/3}\left(\frac{3h^2}{4\pi em}\right) \\
&= \left(\frac{N}{V}\right)^{5/3}\left(\frac{h^2}{2\pi em}\right).
\end{aligned}
\tag{6.30}
$$

This pressure P_0 is often called the *degeneracy pressure*. In this context the adjective *degeneracy* simply means "relatively high occupancy" or "in the quantum limit." The phrase *ground state pressure* is more descriptive. A non-vanishing ground state energy E_0, a non-vanishing ground state pressure P_0, and the vanishing of the heat capacity in the $T \to 0$ limit are outstanding quantum features of the ideal Fermi gas.

6.4 Average energy approximation

The reader who consults texts listed in Appendix VI will find that the usual way of describing ideal Fermi gas is via the occupation number method according to which the entropy is maximized over occupation number macrostates while constrained to conserve the system energy E and particle number N. This approach has become standard and produces an expression for the probability $p_j\left[= e^{-\varepsilon_j/kT}/Z_1\right]$ that a particle of ideal gas will occupy a single-particle microstate whose energy is ε_j. Consequently, procedures for determining the single-particle partition function Z_1 and the quantized energies ε_j are required. Then the entropy of the gas is determined from the Gibbs entropy formula $S = -kN\sum_j p_j \ln p_j$.

In contrast, we have, in Section 6.3, employed and will, in Sections 7.4 and 7.5, employ the average energy approximation according to which all the particles of an ideal gas have the same energy: the average energy of a particle E/N. Then the system entropy $S[= k\ln\Omega]$ is determined through its multiplicity Ω by the number of ideal gas particles N, the number of single-particle microstates n those particles may occupy, and the kind of indistinguishable particles, fermions or bosons, that compose the gas. The number of single-particle microstates n proceeds from quantizing phase space into Plank's constant-sized chunks and depends, according to (6.13), on the system volume V and average energy per particle E/N. In this way an entropy function $S(E,V,N)$ is produced that encapsulates all that can be known of the thermodynamics of a quantum ideal gas.

These two methods, one based on maximizing the entropy over occupation number macrostates and the other based on the average energy approximation, produce many of the same results. Both produce the Sackur–Tetrode entropy in the high temperature, low density, classical limit and both produce important quantum features: third law compliant entropy and heat capacity and the non-vanishing ground state energy and pressure of an ideal Fermi gas. But there are differences. In particular, the ground state energy (6.23) and pressure (6.30) of an ideal Fermi gas predicted by the average energy approximation are a factor of 1.2 higher than those produced by standard methods. Quantities determined by the derivatives of the entropy, like the heat capacity C_V, differ even in their functionality.

There is no doubt that the standard approach to the ideal quantum gases incorporates more physics and is more accurate than the approach of the average energy approximation. However, we have employed the average energy approximation because it generates important results with a minimum of mathematics and directly exploits the properties of the entropy function $S(E,V,N)$.

Example 6.1 conduction electrons, nuclei, and white dwarfs

The conduction electrons within a typical metal such as copper form an ideal Fermi gas that is essentially ground state at room temperature. Since each copper atom contributes approximately one electron to its gas of conducting electrons, the density of copper atoms, $N/V \approx 8.48 \times 10^{28}$ atoms/m^3, is the density of its conduction electrons. Given T = 300 K for room temperature, the normalized temperature (6.29) becomes a number, 1.15×10^{-12}, that is well below 1. Thus the conduction electrons in a metal contribute per electron much less to the heat capacity than that classically expected, $3k/2[= 3R/2N_A]$.

The ground state pressure P_0 can be quite large for the conduction electrons in metals – 46.2 billion N/m^2 for copper at standard density – but is effectively neutralized by the electrostatic attraction between the metallic ions and the conduction electrons of the ideal Fermi gas.

The nucleons (protons and neutrons) within a nucleus also form an ideal Fermi gas. The ground state pressure within a nucleus opposes the strong-force attraction between nucleons and, in this way, provides for the stability of the nucleus and determines its size. A similar balancing occurs between the outward pressure of the electrons of white dwarf stars, stars that at the end of their nuclear fuel cycle have collapsed into objects a million times more dense than the Sun, and the inward force of gravity.

Problems

6.1 Extensivity

Show that the entropy of an ideal Fermi gas is an extensive function of its fluid variables, that is, show that the function $S(E,V,N)$ implied by (6.15) and (6.17) has the property $\lambda S(E,V,N) = S(\lambda E, \lambda V, \lambda N)$.

6.2 De Broglie wavelength

Another way to indicate when an ideal gas enters the quantum regime is to indicate when the de Broglie wavelength $\lambda\ [= h/p]$ of particles with average energy E/N becomes larger than the average spacing $(V/N)^{1/3}$ between particle centers. (a) Derive an inequality that expresses this condition in terms of the average particle energy E/N, the particle density N/V, and universal constants. (b) What values of the occupancy N/n does this condition imply?

6.3 Heat capacity

(a) Derive the heat capacity of an ideal Fermi gas (6.27), that is, derive

$$\frac{C_V}{Nk} = \frac{3(n/N)\ln(1-N/n)}{\left[1+3(N/n)/\{(1-N/n)\ln(1-N/n)\}\right]}$$

from $C_V = T(\partial S/\partial T)_V$. (b) Show that this heat capacity recovers the result expected for a classical, ideal, monatomic gas when $N/n \ll 1$.

6.4 Laser fusion

One method of creating the temperatures and densities necessary for nuclear fusion is to shine laser beams on a pellet of the isotopes of hydrogen: deuterium and tritium. The result sought is a temperature of at least 1 keV and an electron density of about $N/V = 10^{33}/\text{m}^3$ in the imploded core of the fuel pellet. Determine the size of the normalized temperature NkT/E_0 that corresponds to these parameters. Is the imploded core of the fuel pellet in the quantum regime?

6.5 White dwarfs

When a star consumes all its nuclear fuel it no longer generates enough energy to maintain its size and the star will collapse into a compact object, either a white dwarf, a neutron star, or a black hole, depending upon its mass. The ground state or degeneracy pressure of its electron gas is the mechanism that keeps white dwarf stars with masses up to 1.4 times the mass of the sun from further gravitational collapse. Consider that a spherical, homogeneous white dwarf of mass M and radius R will have a negative gravitational potential energy of $U_g = -3GM^2/5R$, and that the inward gravitational pressure exerted on the surface of the white dwarf is $P_g = -(\partial U_g/\partial V)$. A white dwarf is hot enough to be a fully ionized plasma and dense enough to be in the fully quantum regime. Therefore, its electron pressure is $P_0 = (N/V)^{5/3}(h^2/2\pi em)$, and the condition $P_g + P_0 = 0$ establishes the equilibrium of a white dwarf. Assume that a white dwarf is composed of equal numbers of electrons with mass m and protons with mass m_p. Determine how the radius R of a white dwarf depends on its mass M. (Hint: You will need to use the relationship between the volume V and the radius R of a sphere.)

7

Entropy of systems of bosons

7.1 Photons

In 1900 Max Planck formulated an expression for the spectral energy density of blackbody radiation,

$$\rho(v) = \frac{8\pi v^2}{c^3} \frac{hv}{\left(e^{hv/kT} - 1\right)}, \tag{7.1}$$

that successfully reproduced the energy density of blackbody radiation contained within the differential frequency interval v to $v + dv$. But the physical principles behind this formula remained obscure until illuminated during the next 25 years by Planck himself, by Einstein, and by Bose. Bose, in particular, composed all the principles behind Planck's formula (absolute entropy, determinate phase space cells, quantized energy, and particle-like light quanta or photons) into a coherent and fully quantum theory of blackbody radiation.

Photons

The idea of a light quanta or *photon*, introduced by Einstein in 1905, was at first resisted. After all, the wave theory of light had been a tremendous success. But the purpose of the photon concept was to make sense of a few phenomena that wave theory failed to explain. Among these are the *photoelectric effect* in which sufficiently high frequency light shining on a metallic surface ejects electrons from that surface and the *Compton effect* in which X-rays collide with free or essentially free electrons. In these effects, illustrated in Figure 7.1, a photon transfers some energy and momentum to an electron.

Figure 7.1 (a) The photoelectric effect: an ultraviolet photon strikes a metallic surface and ejects an electron. (b) The Compton effect: an X-ray photon scatters from a free electron.

Complementarity

That electromagnetic radiation sometimes behaves as if composed of waves and sometimes behaves as if composed of particle-like photons means that these very different models *complement* each other. Alternatively, electromagnetic radiation has a dual nature that is expressed quantitatively in the relation between the frequency v of the wave and the energy ε and momentum p of the associated photon,

$$\varepsilon = hv \tag{7.2}$$

and

$$p = hv/c. \tag{7.3}$$

Interestingly, Einstein remained skeptical of his own invention, the photon, and of all quantum concepts. Einstein believed that the photon is a mere heuristic, that is, a suggestive yet provisional device, that some day will be replaced with a more fundamental theory.

7.2 Blackbody radiation

Maxwell's equations are linear in the electric and magnetic fields. For this reason electromagnetic waves, and therefore photons, do not interact with one another but achieve equilibrium only by being absorbed and emitted, say, by the walls that form the left chamber in Figure 7.2. These walls are at temperature T. Also in Figure 7.2 a filter allows radiation within the frequency interval from v to $v + dv$ to pass from the left chamber with absorbing and emitting walls to the right chamber with perfectly reflecting walls. In this way the right chamber isolates a monochromatic or single-color part of the blackbody radiation.

We refer to the system of monochromatic radiation composed of photons associated with frequency v as *the subsystem* throughout this section. The

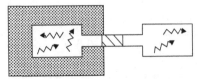

Figure 7.2 Left chamber: blackbody radiation in equilibrium with its material walls at temperature T. Right chamber: monochromatic blackbody radiation surrounded by perfectly reflecting walls. Center: a filter that allows only single-frequency radiation to pass from left to right.

photons of this subsystem have energy hv and momentum hv/c and are iso-tropically directed, that is, directed uniformly in all directions. We adopt the following notation: E^v for the subsystem energy, S^v for its entropy, P^v for its pressure, Ω^v for its multiplicity, $N_p^v \left[= E^v/hv \right]$ for the number of subsystem photons, and N_c^v for the number of states that a single photon with frequency v may occupy. These quantities are related by $S^v = k \ln \Omega^v$, $1/T = \left(\partial S^v / E^v \right)_v$, and $P^v/T = \left(\partial S^v / \partial V \right)_{E^v}$.

Note that all quantities superscripted with v are differentials associated with the subsystem frequency interval v to $v + dv$. For instance, the energy of the subsystem is

$$E^v = V \rho(v) dv \qquad (7.4)$$

where $\rho(v)$ is the spectral energy density. The number of possible photon states N_c^v for photons with momentum $p = hv/c$ is given by

$$N_c^v = \frac{V}{h^3} (\text{momentum space volume})(\text{number of photon polarity states})$$

$$= \frac{V}{h^3} \left(4\pi p^2 dp \right)(2)$$

$$= \frac{V 8\pi p^2 dp}{h^3}$$

$$= \frac{V 8\pi v^2 dv}{c^3}, \qquad (7.5)$$

since the accessible momentum space is a shell in momentum space of radius p and thickness dp and each photon has two possible polarizations.

The photon multiplicity Ω^v is the number of distinct arrangements of N_p^v identical photons among these N_c^v states. Recall the structurally identical problem, discussed in Section 4.4 and illustrated in Figure 4.3, of finding the number of distinct arrangements of identical phonons among oscillator degrees of freedom. Thus, the subsystem multiplicity is

$$\Omega^{\nu} = \frac{\left(N_p^{\nu} + N_c^{\nu} - 1\right)!}{N_p^{\nu}!\left(N_c^{\nu} - 1\right)!} \tag{7.6}$$

and therefore the absolute entropy of the subsystem is

$$S^{\nu} = k\left[\left(N_p^{\nu} + N_c^{\nu}\right)\ln\left(N_p^{\nu} + N_c^{\nu}\right) - N_p^{\nu}\ln N_p^{\nu} - N_c^{\nu}\ln N_c^{\nu}\right], \tag{7.7}$$

where we have assumed $N_p^{\nu} \gg 1$ and $N_c^{\nu} \gg 1$.

In developing the subsystem energy equation of state from $1/T = \left(\partial S^{\nu}/\partial E^{\nu}\right)_V$ it is most convenient to start with (7.7), to keep the numbers of subsystem photons $N_p^{\nu}\left[= E^{\nu}/h\nu\right]$ and subsystem cells $N_c^{\nu}\left[= V8\pi\nu^2 d\nu/c^3\right]$, as parameters, and to proceed as follows:

$$\frac{1}{kT} = \left(\frac{\partial N_p^{\nu}}{\partial E^{\nu}}\right)\left[\ln\left(N_p^{\nu} + N_c^{\nu}\right) + 1 - \ln N_p^{\nu} - 1\right]$$

$$= \left(\frac{1}{h\nu}\right)\ln\left(1 + \frac{N_c^{\nu}}{N_p^{\nu}}\right) \tag{7.8}$$

$$= \left(\frac{1}{h\nu}\right)\ln\left(1 + \frac{V8\pi h\nu^3 d\nu}{c^3 E^{\nu}}\right).$$

Solving (7.8) for $E^{\nu}/Vd\nu \left[= \rho(\nu)\right]$ produces Planck's expression,

$$\rho(\nu) = \frac{8\pi\nu^2}{c^3} \frac{h\nu}{\left(e^{h\nu/kT} - 1\right)} \tag{7.9}$$

for the spectral energy density of blackbody radiation.

Example 7.1 Stefan–Boltzmann law and the radiation constant

Problem: Derive the Stefan–Boltzmann law $E/V = aT^4$ from the spectral energy density of blackbody radiation and express the radiation constant a, first identified in Section 1.9, in terms of fundamental constants.

Solution: Since the subsystem energy E^{ν} is a differential associated with monochromatic radiation in the interval ν to $\nu + d\nu$ and the energy of a system is additive over its parts, the energy E of the complete spectrum of black-body radiation is the integral of E^v over all frequencies from $v = 0$ to $v = \infty$. In particular,

$$E = \int_0^\infty E^v$$
$$= V \int_0^\infty \rho(v)\,dv.$$

Given the spectral energy density (7.9) we find that

$$E = \frac{V8\pi h}{c^3} \int_0^\infty \frac{v^3\,dv}{e^{hv/kT} - 1}$$
$$= \frac{V8\pi k^4 T^4}{c^3 h^3} \int_0^\infty \frac{x^3\,dx}{e^x - 1}$$
$$= \frac{8\pi^5 k^4}{15c^3 h^3} VT^4$$
$$= aVT^4$$

which is the Stefan–Boltzmann law with the radiation constant

$$a = \frac{8\pi^5 k^4}{15c^3 h^3}$$

expressed in terms of fundamental constants.

7.3 Ideal Bose gas

Because we approach the ideal Bose gas in much the same way as we approached the ideal Fermi gas in Section 6.3, we should recall the first few paragraphs of that section. Here we also identify the number of single-particle microstates or cells available to a particle of ideal gas with energy E/N, whether distinguishable, fermion, or boson by

$$n = V \left(\frac{E}{N}\right)^{3/2} \left(\frac{4\pi em}{3h^2}\right)^{3/2}, \tag{7.10}$$

where E, V, and N are the extensive fluid variables that describe the system, h is Planck's constant, and e is the base of the natural logarithm. And we again make use of the average occupation number per cell, that is, the occupancy

$$\frac{N}{n} = \left(\frac{N}{V}\right)\left(\frac{N}{E}\right)^{3/2}\left(\frac{3h^2}{4\pi em}\right)^{3/2}. \tag{7.11}$$

Because any number of bosons may occupy the same single particle cell,

$$0 < \frac{N}{n} \le N \tag{7.12}$$

defines the domain of the occupancy of an ideal Bose gas. And we again adopt the average energy approximation.

The multiplicity of an ideal Bose gas

The number of distinct ways of placing N identical bosons each with energy E/N in n distinct cells is the multiplicity

$$\Omega(E,V,N) = \frac{(N+n-1)!}{N!(n-1)!}. \tag{7.13}$$

The counting that produces (7.13) is structurally identical to counting the number of distinct ways of ordering a collection of N identical balls and $n-1$ identical white dividers. When placed in a row, the $n-1$ dividers separate the N balls into n ordered (and so distinct) groups or cells. See, for comparison, Figure 4.3. Also, recall that we used the combinatorial formula (7.13) when counting phonon microstates in Section 4.4 and when counting photon microstates in Section 7.2 – both examples of counting bosons.

Paul Ehrenfest (1880–1933) questioned Einstein about replacing the multiplicity n^N of an ideal classical gas of distinguishable particles with the multiplicity $(N+n-1)!/[(n-1)!N!]$ of an ideal gas of indistinguishable bosons. For $\Omega = n^N$ implies that the particles of an ideal classical gas occupy their position in single-particle phase space independently of one another. After all, if $\Omega = n^N$ then the system entropy $S [= Nk\ln n]$ is simply the sum of the entropies of the individual particles. However, if $\Omega = (N+n-1)!/[(n-1)!N!]$, no such decomposition of the multiplicity and no such interpretation of the entropy is possible. Apparently, the particles of an ideal Bose gas do not occupy their positions in phase space independently of one another. While Einstein did not object to this characterization he found it inexplicable. That the particles of an ideal Bose gas, and likewise the particles of an ideal Fermi gas, are correlated with one another even in the absence of interparticle forces is a purely quantum effect.

The entropy

Given (7.13) the entropy of an N-particle ideal Bose gas is given by

$$S(E,V,N) = k \ln \left[\frac{(N+n-1)!}{N!(n-1)!} \right]. \tag{7.14}$$

We assume that $N \gg 1$ and $n \gg 1$ in order to deploy Stirling's approximation and so from (7.14) generate

$$\frac{S(E,V,N)}{Nk} = \left(1 + \frac{n}{N}\right) \ln(N+n) - \ln N - \left(\frac{n}{N}\right) \ln n$$
$$= \left(\frac{n}{N}\right) \ln\left(1 + \frac{N}{n}\right) + \ln\left(1 + \frac{n}{N}\right). \tag{7.15}$$

Note that the entropy (7.15) of an ideal Bose gas is, given (7.11), an extensive function of its extensive variables E, V, and N. While Eqs. (7.11), (7.14), and (7.15) describe all there is to know of the thermodynamic behavior of an ideal Bose gas, the particular properties of this gas remain to be explored.

Low occupancy regime

When the occupancy N/n is low, that is, when

$$\frac{N}{n} \ll 1, \tag{7.16}$$

the leading-order terms in an expansion of the entropy (7.15) are

$$\frac{S(E,V,N)}{Nk} = 1 + \ln\left(\frac{n}{N}\right)$$
$$= 1 + \ln\left[\left(\frac{V}{N}\right)\left(\frac{E}{N}\right)^{3/2}\left(\frac{4me\pi}{3h^2}\right)^{3/2}\right] \tag{7.17}$$
$$= \frac{5}{2} + \ln\left[\left(\frac{V}{N}\right)\left(\frac{E}{N}\right)^{3/2}\left(\frac{4m\pi}{3h^2}\right)^{3/2}\right],$$

where in the first step we have used (7.11). This result (7.17) is the Sackur–Tetrode entropy (4.2) of an ideal gas.

Equations of state

The entropy S as given by (7.15) and the occupancy N/n as defined by (7.11) completely describe the thermodynamics of the ideal Bose gas as long as $N \gg 1$ and $n \gg 1$. In particular, its equations of state follow from (7.11), (7.15), $1/T = (\partial S/\partial E)_{V,N}$, and $P/T = (\partial S/\partial V)_{E,N}$. From these we find that

$$\frac{E}{NkT} = \left(\frac{3}{2}\right)\left(\frac{n}{N}\right)\ln\left(1+\frac{N}{n}\right) \tag{7.18}$$

and

$$\frac{PV}{NkT} = \left(\frac{n}{N}\right)\ln\left(1+\frac{N}{n}\right). \tag{7.19}$$

An immediate consequence of the equations of state (7.18) and (7.19) is that

$$E = \frac{3}{2}PV. \tag{7.20}$$

Note that Planck's constant h survives in these equations of state through its appearance in the occupancy (7.11) and that the low occupancy limit reduces the equations of state (7.18) and (7.19) to the familiar ideal gas equations of state $PV = NkT$ and $E = 3NkT/2$.

Low temperature regime

How does the ideal Bose gas behave in the low temperature regime? Expression (7.11) insures that the high occupancy regime is equivalent to the low energy regime while (7.18) insures that the low energy regime is equivalent to the low temperature regime. Therefore, low temperature means high occupancy for the ideal Bose gas just as it did for the ideal Fermi gas.

We are particularly concerned to know if the ideal Bose gas described by the equations of state (7.18) and (7.19) observes the third law. Since we cannot solve (7.11), (7.15), and (7.18) analytically for the function $S(T)$, we again use the occupancy N/n as a parameter that numerically links normalized values of the entropy S and temperature T for constant values of N and V. We plot the resulting function $S(T)$ in Figure 7.3. However, the parameters with which we normalize the temperature of the ideal Bose gas cannot be those used to normalize the temperature of the ideal Fermi gas. After all, the two gases have very different ground states.

A particular ground state of the ideal Bose gas is one in which all N bosons occupy the same single-particle state, and so $n = 1$, and yet its volume remains at V. We find that the energy of this special ground state, denoted E_{00}, is a convenient normalizing energy. In this case (7.10) reduces to $1 = V\left(E_{00}/N\right)^{3/2}\left(4\pi em/3h^2\right)^{3/2}$. Therefore,

$$E_{00} = \frac{N}{V^{2/3}}\left(\frac{3h^2}{4\pi em}\right) \qquad (7.21)$$

is the energy of this ground state in terms of which (7.10) may be expressed as

$$\frac{E}{N^{2/3}E_{00}} = \left(\frac{n}{N}\right)^{2/3}. \qquad (7.22)$$

Using (7.22) to eliminate the energy E from the energy equation of state (7.18) produces a parametric equation,

$$\frac{kTN^{1/3}}{E_{00}} = \left(\frac{2}{3}\right)\frac{(N/n)^{1/3}}{\ln(1+N/n)}, \qquad (7.23)$$

for the normalized temperature $kTN^{1/3}/E_{00}$. This result (7.23) and the normalized entropy S/Nk given by (7.15) allow us to numerically generate coordinate values of S/Nk and $kTN^{1/3}/E_{00}$ and plot the result in Figure 7.3. The occupancies that generate Figure 7.3 range from $N/n = 1$ (upper right) to $N/n = 100$ (lower left center).

Note the surprising shape of $S(T)$. The $S(T)$ curve bends around and forms a nose near $N/n = 16$ at which point the slope $(\partial S/\partial T)_V$ diverges to infinity. These features signal instability. But before we investigate this instability, we consider in more detail the special $n = 1$ ground state in which all the system bosons occupy the same single-particle microstate.

Ground state

It is important to realize that the $n = 1$ ground state is not described by the entropy (7.15) nor by the equations of state, (7.18) and (7.19), that follow from it, nor by any equations that follow from these. After all, these equations result from applying Stirling's approximation to $S = k\ln\Omega$ and so assume $n \gg 1$ as well as $N \gg 1$. In describing the $n = 1$ ground state we depend solely on (7.11)–(7.13) and the combinatoric expression for entropy (7.14). Therefore, the $n = 1$ ground state multiplicity is $\Omega = 1$, and consequently the ground state entropy $S = 0$. Furthermore, this ground state is distinct from the macrostates illustrated

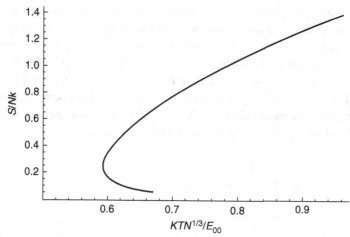

Figure 7.3 The normalized entropy S/Nk versus the normalized temperature $kTN^{1/3}/E_{00}$ when N and V are constants for occupancies ranging from $N/n = 1$ (upper right) to $N/n = 100$ (lower left-center) as determined from (7.15) and (7.23).

in Figure 7.3. In principle, the $n = 1$ ground state with energy denoted E_0 is distinct from the normalizing ground state energy E_{00} because the $n = 1$ ground state volume V_0 can be different from the volume V of the $n \gg 1$ macrostates.

Thermodynamic instability

Because the fundamental equation $dE = T\,dS - P\,dV$ implies that the heat capacity at constant volume $C_V \left[= (\partial E/\partial T)_V \right]$ is given by $T(\partial S/\partial T)_V$, a negative slope $(\partial S/\partial T)_V$ indicates a negative heat capacity. And a system with negative heat capacity cannot be in equilibrium with a constant-temperature environment. After all, the temperature of a system with negative heat capacity would decrease when the system was heated and increase when cooled. Either behavior would drive the system further from equilibrium with a constant-temperature environment. Therefore, according to the calculation displayed in Figure 7.3, the ideal Bose gas is unstable for occupancies greater than about $N/n \approx 16$.

Actually, the ideal Bose gas is thermodynamically unstable for much smaller occupancies. To see this, consider that an isolated system is thermodynamically unstable whenever it can transition to another macrostate with higher entropy.

In actual fact experimental realizations of low temperature ideal Bose gases
are never isolated and a path always exists for the system to lose energy and
occupy its ground state. While experiments vary, typically the gas is in contact
with an environment at constant temperature T and constant pressure P as T is
slowly decreased. In this case, the ideal Bose gas will begin to occupy its $n = 1$
ground state when by doing so it can decrease its Gibbs free energy. In particular, the Gibbs free energy $G\left[= E - TS + PV\right]$ of the $n > 1$ state begins to exceed
the Gibbs free energy of the $n = 1$ ground state whenever

$$G_{n>1} = G_{n=1},$$
$$E - TS + PV = E_0 - TS_0 + PV_0,$$
$$\frac{5}{3}E - TS = 0,$$
$$kT\left(\frac{5}{3}\right)\left(\frac{3}{2}\right)\left(\frac{n}{N}\right)\ln\left(1+\frac{N}{n}\right) = kT\left[\ln\left(1+\frac{n}{N}\right)-\left(\frac{n}{N}\right)\ln\left(1+\frac{N}{n}\right)\right],$$
$$\left(\frac{3}{2}\right)\left(\frac{n}{N}\right)\ln\left(1+\frac{N}{n}\right) = \ln\left(1+\frac{n}{N}\right),$$

(7.24)

where in the second step of this equation sequence we use (7.20) on the left-
hand side and $S_0 = 0$ on the right hand side and recognize that the ground state
energy and volume, E_0 and V_0, are ignorably small compared with E and V.

The critical occupancy that solves (7.24) is

$$\frac{N}{n_c} = 0.392. \tag{7.25}$$

Whenever $G_{n>1} > G_{n=1}$ the occupancy $N/n > N/n_c$ and the ideal Bose gas
becomes thermodynamically unstable. Alternatively, whenever $N/n < N/n_c$
the ideal Bose gas is thermodynamically stable. This critical occupancy N/n_c
separating stability from instability corresponds, according to (7.23), to a normalized critical temperature $kT_cN^{1/3}/E_{00} = 1.48$ or, equivalently, to

$$kT_c = 1.48\frac{E_{00}}{N^{1/3}}$$
$$= 0.130\left(\frac{N}{V}\right)^{2/3}\left(\frac{h^2}{m}\right),$$

(7.26)

a result that is a factor of 1.5 greater than the critical temperature predicted
by models that incorporate more physics. The critical point, for which

$kT_cN^{1/3}/E_{00} = 1.48$, $S_c/Nk = 2.11$, and $N/n_c = 0.392$, is on the curve shown in Figure 7.3 but far beyond the figure's upper right hand corner.

7.4 Bose–Einstein condensate

If we cool an ideal Bose gas to a temperature just below its critical temperature T_c and, consequently, to an occupancy just above its critical occupancy N/n_c [$= 0.392$], the gas will break into two parts with each part being expelled to a nearby relative minimum in the Gibbs free energy, that is, to a stable region of thermodynamic phase space. When the temperature of the gas remains close to its critical temperature, the largest share of the particles will remain at the critical occupancy N/n_c and the remaining, smaller share will condense into the $n = 1$, $S = 0$ ground state. This transition is called a phase transition and the material in each share is a called a different phase of the gas. Thus, when $T < T_c$ the system becomes a composite, two-phase system. The phase remaining at the onset of instability is a normal, if low temperature, ideal Bose gas while the phase in the ground state is one whose macrostate consists of one single-particle microstate.

Albert Einstein first predicted the existence of the $n = 1$ phase or condensate early in 1925. Einstein, who was inspired by Bose's 1924 analysis of blackbody radiation, had applied Bose's counting method, that is, Bose's statistics, to an ideal gas composed of identical material particles. Then, on the basis of an extended analysis, Einstein predicted the existence of this condensed, $n = 1$, single-particle microstate phase we now called the *Bose–Einstein condensate*.

Two-phase regime

As the system temperature T drops below its critical temperature T_c the number of particles occupying the uncondensed phase N_{un} decreases and the number occupying the $n = 1$ ground state phase N_0 increases in such a way that the total number of particles

$$N = N_{un} + N_0 \qquad (7.27)$$

is conserved. Alternatively, the fraction of particles in the uncondensed phase N_{un}/N decreases and the fraction of particles in the ground state phase N_0/N increases as the temperature T drops further below T_c. The system energy and entropy are, however, not conserved. Both decrease as the system temperature T decreases to below T_c.

Because the gas particles always seek the closest stable macrostate consistent with its current temperature T, that is, seek the state with the smallest Gibbs free energy, the N_{un} uncondensed particles remain at critical occupancy so that

$$\frac{N_{un}}{n_{un}} = \frac{N}{n_c}[= 0.392] \tag{7.28}$$

as both N_{un} and n_{un} decrease. Thus, condition (7.28) expresses the stability of the particles in the uncondensed phase.

This stability relation (7.28) has important consequences. Using the definition of the occupancy (7.11) to expand both sides of (7.28) we find

$$\frac{N_{un}}{n_{un}} = \frac{N}{n_c}$$

$$\left(\frac{N_{un}}{V}\right)\left(\frac{N_{un}}{E_{un}}\right)^{3/2}\left(\frac{3h^2}{4\pi em}\right)^{3/2} = \left(\frac{N}{V}\right)\left(\frac{N}{E_c}\right)^{3/2}\left(\frac{3h^2}{4\pi em}\right)^{3/2} \tag{7.29}$$

$$\frac{N_{un}^{5/2}}{E_{un}^{3/2}} = \frac{N^{5/2}}{E_c^{3/2}}$$

where E_{un} is the energy of those particles that remain in the uncondensed, critical state. Similarly we compare the energy equation of state (7.18) when $T = T_c$ and, consequently, all the particles are at the critical point,

$$\frac{E_c}{NkT_c} = \frac{3}{2}\left(\frac{n_c}{N}\right)\ln\left(1 + \frac{N}{n_c}\right), \tag{7.30}$$

with the energy equation of state (7.18) when $T < T_c$ and only part of the uncondensed particles are at the critical point,

$$\frac{E_{un}}{N_{un}kT} = \frac{3}{2}\left(\frac{n_{un}}{N_{un}}\right)\ln\left(1 + \frac{N_{un}}{n_{un}}\right). \tag{7.31}$$

Given the stability condition (7.28) the right-hand sides of Eqs. (7.30) and (7.31) are equal. Therefore,

$$\frac{E_{un}}{N_{un}T} = \frac{E_c}{NT_c} \tag{7.32}$$

when $T < T_c$. Eliminating the ratio E_{un}/E_c from (7.29) and (7.32) produces

$$\frac{N_{un}}{N} = \left(\frac{T}{T_c}\right)^{3/2}, \tag{7.33}$$

which, given particle conservation (7.27), is equivalent to

$$\frac{N_o}{N} = 1 - \left(\frac{T}{T_c}\right)^{3/2}, \tag{7.34}$$

an expression that describes the way in which the ground state is populated as the temperature T drops below the critical temperature T_c and approaches absolute zero. According to (7.34), when $T = T_c$ all the particles are in the critical, uncondensed phase and when $T = 0$ all the particles are in the condensed, ground state phase.

The entropy of the composite, two-phase system is the sum of the entropies of its two phases. Since the entropy of the ground state phase vanishes, we have for the entropy of the composite system

$$
\begin{aligned}
S &= N_{un}k\left\{\ln\left[1+\frac{n_{un}}{N_{un}}\right]+\left(\frac{n_{un}}{N_{un}}\right)\ln\left[1+\frac{N_{un}}{n_{un}}\right]\right\}, \\
&= N_u k\left\{\ln\left[1+\frac{n_c}{N}\right]+\left(\frac{n_c}{N}\right)\ln\left[1+\frac{N}{n_c}\right]\right\} \\
&= 2.11 N_{un}k,
\end{aligned}
\tag{7.35}
$$

where in the first and second steps we have used the stability condition (7.28). Given (7.35) and (7.33), the normalized entropy of the two-phase system is

$$
\begin{aligned}
\frac{S}{Nk} &= 2.11\frac{N_{un}}{N} \\
&= 2.11\left(\frac{T}{T_c}\right)^{3/2}.
\end{aligned}
\tag{7.36}
$$

Figure 7.4 replots Figure 7.3 with the unstable part of the $S-T$ relation subtracted and the two-phase part of the $S-T$ relation (7.36) added. The solid curve shows the entropy. Since $S \to 0$ as $T \to 0$, this composite system observes the third law. Also shown (dashed curve) is the normalized heat capacity C_V/Nk. (See Problem 7.7.) Because the entropy is continuous at the critical temperature while its first derivative with respect to temperature $(\partial S/\partial T)_V \left[= C_V/T\right]$ is discontinuous, this transition, at least in the average energy approximation, is a *first-order* phase transition.

Albert Einstein remarked in a letter to Paul Ehrenfest, "The theory [of the ideal Bose gas] is pretty, but is there some truth to it?" The world had to wait some years for experimental verification of Einstein's prediction of a

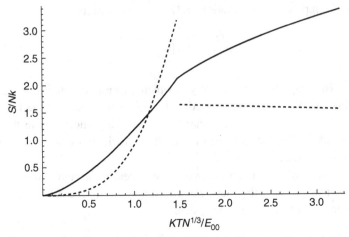

Figure 7.4 Normalized entropy S/Nk (solid curve) versus the normalized temperature $kTN^{1/3}/E_{00}$ for densities ranging from $N/n = 0.3$ (upper right) to the $n = 1$ ground state (lower left). Determined from (7.15) and (7.23) when $T \geq T_c$ and from (7.23) and (7.36) when $T \leq T_c$. Also shown is the normalized heat capacity $C_V = T(\partial S/\partial T)_V$ (dashed curve) versus normalized temperature as taken from Problem 7.7.

low temperature condensed phase of an ideal Bose gas. In 1938 Fritz London interpreted the phase transition between helium I and helium II as a transition to a Bose–Einstein condensate. The observed critical temperature was lower than predicted, but helium atoms attract each other at the liquid densities at which the condensation was observed and these attractive interatomic forces would explain this discrepancy. It was not until 1995 that Eric Cornell and Carl Wieman of the University of Colorado cooled a gas of weakly interacting bosons (rubidium atoms) below the critical temperature and produced a Bose–Einstein condensate. For this work they, along with Wolfgang Ketterle for shortly thereafter creating a Bose–Einstein condensate with sodium atoms at the Massachusetts Institute of Technology, received the 2001 Nobel Prize in Physics.

7.5 Modeling the ideal gas

We reflect on our derivations of the ideal Fermi and ideal Bose gases in Sections 6.3 and 7.3. The quantum properties incorporated into these models include: phase space cells determined by Planck's constant h, the absolute entropy $S\left[= k\ln\Omega\right]$, the indistinguishability of identical particles, and the particular ways we count boson and fermion microstates. Thus, these models do not depend upon quantizing dynamical properties like the particle energy or momentum nor on assuming a particular shaped volume that makes such quantizing convenient.

Average energy approximation

We also used the average energy approximation in describing the ideal Fermi and ideal Bose gases. The average energy approximation shifts the ground state energy and pressure of the ideal Fermi gas by a factor of approximately 1.2 and the critical temperature of the ideal Bose gas by a factor of approximately 1.5 above values predicted by more complete, standard derivations. Furthermore, the average energy approximation distorts the derivatives of the entropy function in the low temperature, quantum regime. Nevertheless the average energy approximation allows us to quickly reproduce the distinguishing features of the ideal quantum gases: entropies and heat capacities that vanish in the $T \to 0$ limit, non-vanishing ground state energy and pressure of the ideal Fermi gas, and a critical temperature below which the ideal Bose gas becomes two phase.

Quantum and classical descriptions

In summarizing the differences between the ideal classical and ideal quantum gases we place those properties associated with a quantum description in the left column and those properties associated with a classical description in the right column. Note that we include extensivity in both columns and that consistency with the Sackur–Tetrode entropy in the semi-classical or low occupancy limit is the only property that applies exclusively to ideal gases.

Quantum description	Classical description
Entropy is extensive	Entropy is extensive
Identical particles are indistinguishable	Identical particles are distinguishable
Entropy is third law compliant	Entropy is not third law compliant
Absolute entropy $S\left[= k\ln\Omega\right]$	Relative entropy $S\left[= c + k\ln\Omega\right]$
Sackur–Tetrode entropy when occupancy $N/n \ll 1$	Consistent with Sackur–Tetrode entropy
H becomes Planck's constant h	H is arbitrary
h survives in equations of state	H vanishes from equations of state

Problems

7.1 Radiation pressure

(a) Derive an expression for the spectral pressure of blackbody radiation P^v in terms of the radiation frequency v, the system temperature T, and universal constants. (Hint: Start with the spectral entropy S^v (7.7), use $P^v/T = (\partial S^v/\partial V)_E$, proceed as in deriving (7.9), and eliminate E^v with $E^v/V\,dv = \rho(v)$.) (b) Integrate P^v from $v = 0$ to $v = \infty$ to get the total pressure P in terms of T and universal constants. (You will need to complete the integral with software or look it up in Appendix II: Formulary.) (c) Show that the pressure P and internal energy E of blackbody radiation are related by $P = E/3V$.

7.2 Radiation pressure at the center of the Sun

The temperature at the center of the Sun is approximately 2×10^7 K. Find the corresponding blackbody radiation pressure. What is the ratio of this radiation pressure to the gas pressure (4×10^{11} atmospheres) at the center of the Sun? (Hint: consult Problem 7.1 or reread Section 1.9.)

7.3 Maximum radiation

The energy of blackbody radiation within the frequency range v to $v + dv$ is $\rho(v)dv$ where $\rho(v)$ is the spectral energy density given by (7.1) or (7.9). (a) Given that the frequency v and wavelength λ of electromagnetic radiation in a vacuum are related by $c = \lambda v$, find the spectral energy density per differential wavelength $\varepsilon(\lambda)$ defined so that $\varepsilon(\lambda)d\lambda = \rho(v)dv$. (b) Find an expression, called Wien's law, for the wavelength λ_{max} that maximizes the spectral energy density per differential wavelength $\varepsilon(\lambda)$. (You will have to numerically solve a simple equation.) (c) Given that the Sun radiates the most energy at a wavelength of close to 5×10^{-7} m, what is the effective temperature of the surface of the Sun?

7.4 Number of photons

Find an expression for the number of blackbody radiation photons as a function of the volume V and temperature T. (Hint: Integrate E^v/hv from $v = 0$ to $v = \infty$. You will have to numerically evaluate the integral $\int_0^\infty x^2/(e^x - 1)dx$ or see the Formulary in Appendix II.)

7.5 Two-phase energy

The context is Sections 7.3 and 7.4 on the Bose–Einstein gas and condensate. Assume that the particles in the condensed state make an ignorably small contribution to the total energy of the system. From (7.18) find an expression the energy E of the composite system in terms of N, T and T_A in the two-phase region, that is, when $T < T_c$.

7.6 Photon condensate?

A cavity full of electromagnetic radiation is a gas of photons, and photons are bosons. Why, then, do photons never condense at low temperatures?

7.7 Heat capacity of ideal Bose gas and condensate

(a) Show that the heat capacity of the ideal Bose gas when $T \geq T_c$ is given by

$$\frac{C_V}{Nk} = \frac{(n/N)(1+N/n)\left[\ln(1+N/n)\right]^2}{N/n-(1/3)(1+N/n)\ln(1+N/n)}.$$

(b) Show that the heat capacity $C_V = T(\partial S/\partial T)_V$ of the composite ideal Bose gas in the two-phase region where $T \leq T_c$ is given by

$$\frac{C_V}{Nk} = 3.16\left(\frac{T}{T_c}\right)^{3/2}.$$

These expressions are used in the construction of the dashed curve in Figure 7.4.

8

Entropy of information

8.1 Messages and message sources

Information technologies are as old as the first recorded messages, but not until the twentieth century did engineers and scientists begin to quantify something they called *information*. Yet the word *information* poorly describes the concept the first information theorists quantified. Of course specialists have every right to select words in common use and give them new meaning. Isaac Newton, for instance, defined *force* and *work* in ways useful in his theory of dynamics. But a well-chosen name is one whose special, technical meaning does not clash with its range of common meanings. Curiously, the *information* of information theory violates this commonsense rule.

Compare the opening phrase of Dickens's *A Tale of Two Cities*: *It was the best of times, it was the worst of times* ... with the sequence of 50 letters, spaces, and a comma: *eon jhktsiwnsho d ri nwfnn ti losabt,tob euffr te* ... taken from the tenth position on the first 50 pages of the same book. To me the first is richly associative; the second means nothing. The first has meaning and form; the second does not. Yet these two phrases could be said to carry the same information content because they have the same source. Each is a sequence of 50 characters taken from English text.

Information theorists and communications engineers are appropriately concerned with the ability to transmit messages of a given form rather than with the content of particular messages. According to Claude Shannon (1916–2001), whose 1948 paper "The Mathematical Theory of Communication" initiated the study of *information theory*,

> These semantic aspects of communication are irrelevant to the engineering problem. The significant aspect is that the actual message is one *selected from a set* of possible messages. The system must be designed to operate for each possible selection, not just the one which will actually be chosen since this is unknown at the time of design.

140

   The set of possible messages or symbols from which an actual message is selected is a *message source* that contains or produces messages. Information theory is concerned with characterizing message sources rather than with characterizing particular messages.

8.2 Hartley's information

Ralph V. L. Hartley (1888–1970) was the first, in 1928, to quantify the information content of a message source with just two numbers: n, the number of characters in the equal-length sequences that compose its messages, and s, the number of equiprobable symbols that each character may assume. Note that the s^n distinct messages in Hartley's message source must themselves be equiprobable since they are composed of equal numbers of equally probable symbols. While Hartley's s symbols could be letters of the alphabet or dots and dashes or the natural numbers, no violence is done his logic if the symbols are themselves equiprobable messages each considered as a unitary whole.

   Hartley required that the measure of *information H* of such a message source have two properties. First, H must be proportional to n, the number of characters in each message sequence. Thus,

$$H = n\, f(s). \tag{8.1}$$

And second, H must be a monotonically increasing function,

$$H = g(s^n), \tag{8.2}$$

of the number s^n of distinct, equiprobable messages the source contains. After all, a message source containing messages with twice as many characters should contain twice the information and a message source containing more messages should contain more information.

   According to requirements (8.1) and (8.2),

$$n\, f(s) = g(s^n) \tag{8.3}$$

where $f(x)$ and $g(x)$ are monotonically increasing functions on the domain $x \geq 0$. The only differentiable, monotonically increasing functions $f(x)$ and $g(x)$ that satisfy (8.3) are

$$\begin{aligned} f(x) &= g(x) \\ &= c \ln x, \end{aligned} \tag{8.4}$$

where c is an arbitrary positive constant. (Problem 8.1 outlines a proof of this statement.) Thus, Hartley defined the information content of a message source

containing messages with n characters each character of which may be realized
with s equally probable symbols as

$$H = c \ln s^n. \tag{8.5}$$

Equation (8.5) neatly illustrates a general principle: *The information content
of a message source is proportional to the logarithm of the number of distinct,
equiprobable messages it contains.*

No experiment can determine the value of the constant c. However, the
constant c can be used to determine when $H = 1$, that is, c can be used to
determine the unit of information. When $c = 1$, $H/n = \ln s$ and H is measured
in *nats* (after *natural* logarithm). Hartley used $c = 1/\ln 10$ in his 1928 paper.
In this case,

$$\frac{H}{n} = \frac{\ln s}{\ln 10} \tag{8.6}$$
$$= \log_{10} s.$$

Thus a message source composed of the numerals 0, 1, 2,... and 9 has $s = 10$
and so is said to have one *Hartley* of information content per message character. The choice $c = 1/\ln 2$ is, for reasons that will soon be apparent, very
convenient. Then

$$\frac{H}{n} = \frac{\ln s}{\ln 2} \tag{8.7}$$
$$= \log_2 s,$$

and H/n is measured in *bits* (a contraction of *binary digits*). A message source
composed of all possible messages of n characters each character of which can
be realized with only 2 [$= s$] equiprobable characters contains 1 bit of information per character. In general the constant c determines the base b of the
logarithm. When $c = 1/\ln b$, $H = \log_b s^n$.

An application

Occasionally one hears that a particular message contains so many bits of
information. For instance, "011100 contains 6 bits of information." To be more
correct one should say that the message 011100 appears to come from a message source composed of 6-character sequences of equally probable zeros and
ones. In this case, $n = 6$, $s = 2$, and, according to (8.5), the information content
of the message source that produced the message 011100 is 6 [$= 6 \log_2 2$] bits,
4.16 [$= 6 \ln 2$] nats, or 1.81 [$= 6 \log_{10} 2$] Hartleys.

The symbol H

Why did Hartley adopt the symbol H for information? We simply do not know. That Hartley wanted to memorialize his own name seems unlikely. He may have thought that $c \ln s^n$ was structurally similar to the H of Boltzmann's celebrated H theorem. But, if so, Hartley was at least partly mistaken. For Hartley's H is, apart from normalizing constants, the negative of Boltzmann's H. Thus, the larger the Boltzmann H of a message source, the smaller its information H.

Example 8.1 Shuffled deck

Problem: A 52-card deck is completely shuffled. (a) What is the information content of the shuffled deck taken as a whole? (b) What is the information content of a 5-card poker hand drawn from the shuffled deck?

Solution: Here we appeal to the general principle that *the information content of a message source is proportional to the logarithm of the number of distinct, equiprobable messages it contains.* Consider the possible arrangements of the shuffled deck and of the poker hand. Since each of these arrangements is equally probable we need only count their number: 52! different equally probable arrangements of the shuffled deck and $52 \times 51 \times 50 \times 49 \times 48$ different equally probable arrangements of the 5-card poker hand. Therefore, the information content of the shuffled deck is $\log_2 52!$ or 226 bits and the information content of the 5-card poker hand is $\log_2 (52 \times 51 \times 50 \times 49 \times 48)$ or 22.6 bits.

8.3 Information and entropy

Hartley's information measure $c \ln s^n$ is comparable not to Boltzmann's H but rather to the absolute entropy $k \ln \Omega$ of an isolated system. The two measures, information and entropy, are similarly constructed. Other comparisons can also be made including:

a symbol	with	a microstate of a simple system
a single-character message	with	a simple system
a character sequence	with	a complex system
a particular message	with	a microstate of a complex system
a message source	with	an ensemble of complex systems
the number of equiprobable messages	with	the number of equiprobable microstates
the constant c	with	Boltzmann's constant k
the information of a message source	with	the entropy of an isolated system

But a deep analogy does not imply identity. And a single name, either *entropy* or *information content* for the two concepts, insufficiently distinguishes between a message source and an ensemble of physical systems. After all, physical entropy is a consequence of the laws of thermodynamics and obeys these laws, while information content is a descriptive measure with certain built-in properties. Apparently the two concepts are substantively distinct if formally similar.

Missing information

A third term, *missing information*, more appropriately references the parallel structures of the entropy of physics and chemistry and the information of information theory. Hartley's information measure $c \ln s^n$ is actually the *missing information* needed to select one n-character message each character of which can be replaced with s symbols out of a message source containing all possible such messages. And if we divide the physical entropy S by Boltzmann's constant k, the result $S/k \, [= \ln \Omega]$ is also the *missing information* needed to specify the microstate of a physical system.

Example 8.2 Missing information

Problem: What is the missing information per atom of a three-dimensional Einstein solid composed of carbon atoms in diamond form at a temperature of 300 K?

Solution: This problem is equivalent to that of finding the value of S/Nk implied by the entropy of the Einstein solid (4.14),

$$S(E) = c(N) + 3Nk \left\{ \left(\frac{E}{3Nhv} + \frac{1}{2} \right) \ln \left(\frac{E}{3Nhv} + \frac{1}{2} \right) \right.$$
$$\left. - \left(\frac{E}{3Nhv} - \frac{1}{2} \right) \ln \left(\frac{E}{3Nhv} - \frac{1}{2} \right) \right\},$$

with its energy (4.13),

$$E = 3Nhr_o \left[\frac{1}{2} + \frac{1}{\left(e^{hv_o/kT} - 1 \right)} \right],$$

for parameters characteristic of diamond ($hv_o / k = 1325 \text{ K}$) at temperature $T = 300$ K. We must also set $c(N)$ equal to zero because only the missing information of a third law compliant physical system is comparable to the information content of a message source. We find that $hv_o / kT = 1325 / 300 = 4.42$, $E / 3Nhr_o = 0.512$, and, therefore, $S/Nk = 0.196$ bits of missing information per carbon atom.

8.4 Shannon entropy

In 1948 Claude Shannon generalized Hartley's information measure and searched for a name to replace the potentially misleading *information*. After considering and rejecting *choice* and *uncertainty*, Shannon settled on *entropy* – no doubt because of the deep analogy between the entropy of physics and chemistry and the information of communications theory. Recall that Hartley's information H $[= c \ln s^n]$ describes the entropy of a message source composed of equally probable messages or symbols. Such a message source is analogous to an ensemble of isolated systems with equally probable microstates. But just as a non-isolated physical system may occupy its states with unequal probabilities, a message source more complicated than Hartley's may be defined that realizes its symbols with unequal probabilities.

Consider a message composed of the 26 letters of the English alphabet. According to Hartley, the information content per character H/n of a message composed of these symbols is $\log_2 26$ or 4.70 bits – assuming the 26 letters of the English alphabet appear with equal probability p $[= 1/26 = 0.0385]$. But a brief examination of English text reveals that the letters of the alphabet do not appear with equal probability. In particular, the letter e appears most frequently, in fact, typically with frequency 0.13. The letter t appears with frequency 0.091, j with frequency 0.015, and z with frequency 0.0074. (See Table 8.1.) If we want to describe a message source that produces text with a frequency of symbols characteristic of English text, we cannot assume that the letters of the alphabet are equally probable.

Constructing the Shannon entropy

Shannon generalized Hartley's information content by constructing an information measure that allows for symbols with unequal probabilities. In reformulating Shannon's argument we seek the missing information or entropy $H(n_1, n_2, ..., n_s)$ of a message source composed of all possible n-character sequences for which $n = n_1 + n_2 + \cdots + n_s$ that, in turn, are composed of s symbols the ith one of which occurs n_i times. For this purpose we consider only very long ($n \gg 1$) equal-length messages in which the ith symbol appears with a frequency n_i/n that closely matches its probability so that

$$p_i = \frac{n_i}{n}, \tag{8.8}$$

where, of course,

$$\sum_{i=1}^{s} n_i = n \tag{8.9}$$

and $\sum_i p_i = 1$. Since each long, equal-length message in the source is equiprobable even if the symbols that compose the messages appear in each message with unequal probabilities, we can apply the general principle suggested by Hartley's analysis: *The missing information or entropy of a message source is proportional to the logarithm of the number of distinct, equiprobable messages it contains.* Since $n!/(n_1!n_2!\cdots n_s!)$ is the number of distinct, equiprobable n-character sequences that can be formed out of n_1 occurrences of the first symbol, n_2 occurrences of the second symbol and so on, we have

$$H(n_1, n_2, \cdots, n_s) = c \ln \left[\frac{n!}{n_1! n_2! \cdots n_s!} \right]. \tag{8.10}$$

Secondly, we require that the missing information or entropy $H(n_1, n_2, \ldots n_s)$ of an n-character sequence be n times the entropy of a one-character sequence $H(p_1, p_2, \ldots, p_3)$ governed by the same probabilities, that is,

$$\begin{aligned} H(n_1, n_2, \ldots, n_s) &= H(np_1, np_2, \ldots, np_s) \\ &= nH(p_1, p_2, \ldots, p_s), \end{aligned} \tag{8.11}$$

where the n_i and the p_i are related by (8.8). Requirement (8.11) is a generalization of Hartley's requirement (8.1) and also is analogous to the requirement that the physical entropy be an extensive function of its variables.

The two requirements (8.10) and (8.11) together imply that

$$\begin{aligned} H(p_1, p_2, \ldots, p_s) &= \frac{1}{n} H(n_1, n_2, \ldots, n_s) \\ &= \frac{c}{n} \ln \left[\frac{n!}{n_1! n_2! \cdots n_s!} \right] \\ &= \frac{c}{n} \left\{ n \ln n - n - \sum_{i=1}^{s} (n_i \ln n_i - n_i) \right\} \\ &= \frac{c}{n} \left\{ n \ln n - \sum_{i=1}^{s} n_i \ln n_i \right\} \\ &= \frac{-c}{n} \sum_{i=1}^{s} n_i \ln \left(\frac{n_i}{n} \right) \\ &= -c \sum_{i=1}^{s} p_i \ln p_i, \end{aligned} \tag{8.12}$$

where the first line of this equation sequence is from (8.11), the second uses (8.10), the third assumes that $n_i \gg 1$, and the last exploits (8.8). The missing information or entropy of a message source composed of one-character messages realized with unequal probabilities,

$$H\left(p_1, p_2, ..., p_s\right) = -c \sum_{i=1}^{s} p_i \ln p_i, \tag{8.13}$$

is called the *Shannon entropy*.

Reduction to expected results

The Shannon entropy (8.13) reduces to expected results in appropriate limits. For instance, when all the symbols are equiprobable so that $p_1 = p_2 = \cdots = p_s = 1/s$, the result

$$H(p_1, p_2, ..., p_s) = -\frac{c}{s} \sum_{i=1}^{s} \ln\left(\frac{1}{s}\right)$$
$$= c \ln s \tag{8.14}$$

recovers Hartley's information content for a one-character message realized with s equiprobable symbols. When one symbol is certain with probability 1 and the others appear with probability 0, the Shannon entropy

$$H(1, 0, \cdots, 0) = -c\left\{1 \times \ln 1 + 0 \times \ln 0 + 0 \times \ln 0 + \cdots\right\},$$
$$= 0 \tag{8.15}$$

where here we stipulate that $0 \times \ln 0 = 0$ in order to recover the limit of $x \ln x$ as $x \to 0$. Thus, a message source that invariably produces the same symbol or message has zero Shannon entropy.

Example 8.3 Two symbols

Problem: A message source produces two symbols, the first with probability p and the second with probability $q\ [= 1 - p]$. What value of p maximizes the Shannon entropy of this message source?

Solution: The Shannon entropy of this one-character, two-symbol source is

$$H(p, q) = -p \log_2 p - q \log_2 q$$
$$= -p \log_2 p - (1 - p) \log_2 (1 - p)$$

when measured in bits. The value of p that maximizes $H(p, 1 - p)$ is the solution of

$$-\log_2 p - 1 + \log_2 (1 - p) + 1 = 0,$$

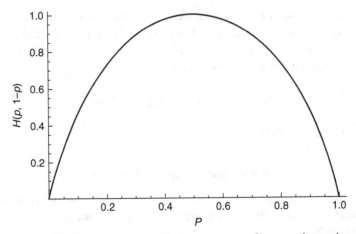

Figure 8.1. The Shannon entropy $H(p, 1-p)$ versus p of a source that produces two symbols with probabilities p and $1-p$. Used in Example 8.3.

that is, $p = 1/2$. The graph of $H(p, 1-p)$ shown in Figure 8.1 confirms that $p = 1/2$ is a maximum. Therefore, the maximum Shannon entropy is $H(1/2, 1/2) = \log_2 2$ or 1 bit.

Example 8.4 Entropy of English text

Problem: What is the Shannon entropy of a source that produces single characters of the alphabet with frequencies observed in English text?

Solution: The frequencies of the letters of the alphabet in English text (displayed in Table 8.1) are easily found online. We assume that these frequencies approximate probabilities. The longer the texts in which the frequencies are observed, the more accurate is this assumption. The calculation of the Shannon entropy per character of English text is then straightforward if tedious. Thus,

$$H(0.0817, 0.0149, \cdots) = -0.0817 \times \log_2(0.0817) - 0.0149 \times \log_2(0.0149) \cdots$$
$$= 4.47 \text{ bits,}$$

a value somewhat smaller than the 4.70 bits per character that would be produced if all 26 letters were equally probable. Apparently, the letters in English text appear with less uncertainty, less missing information, and less Shannon entropy than they would if equally probable.

8.5 Fano code

One application of Shannon entropy is to evaluate the efficiency of coding one alphabet of symbols, in terms of which an original message source is

Table 8.1. *Frequency with which the letters of the alphabet appear in English text.*

Letter	Probability	Letter	Probability	Letter	Probability
a	0.082	j	0.0015	s	0.063
b	0.015	k	0.0077	t	0.091
c	0.028	l	0.040	u	0.028
d	0.043	m	0.024	v	0.0098
e	0.13	n	0.067	w	0.024
f	0.022	o	0.075	x	0.0015
g	0.020	p	0.019	y	0.020
h	0.061	q	0.00095	z	0.00074
i	0.070	r	0.060		

composed, into another alphabet of symbols, typically binary digits, convenient for electronic transmission. The more closely the Shannon entropy of the coded message source matches the Shannon entropy of the original message source, the more efficient the code.

Suppose, for instance, a message source produces the symbols a, b, c, and d with the probabilities 0.5, 0.2, 0.2, and 0.1 as described in Table 8.2. The Shannon entropy per character of this source is

$$-0.5 \times \log_2 0.5 - 2 \times 0.2 \times \log_2 0.2 - 0.1 \times \log_2 0.1 \doteq 1.76 \text{ bits.} \quad (8.16)$$

The coder's task is to translate the letters a, b, c, and d into combinations of transmissible binary digits 0 and 1. Obviously, the most frequently occurring letter, a, should be represented by the shortest sequence of binary digits and so on, but how exactly? *Fano code* is one method. The goal of Fano code is to use $\log_2 (1/p_i)$ binary digits to represent the ith symbol. Since $\log_2 (1/p_i)$ is not always an integer, Fano code can only approximate this goal. The algorithm for Fano coding this particular message source is illustrated in the tree graph of Figure 8.2.

According to the algorithm that produces the tree graph and the Fano code it represents, we arrange the original symbols in order of decreasing probability, $a(0.5)$, $b(0.2)$, $c(0.2)$, $d(0.1)$, on the first leaf of the tree in Figure 8.2. Then we divide this ordered group into two ordered groups with as nearly equal total probability as possible, $a(0.5)$ and $b(0.2)$, $c(0.2)$, and $d(0.1)$, on the second-level leaves. If the symbol is in the leftmost group, its first binary digit is 0; if in the rightmost group, its first binary digit is 1. Each leaf with more than one symbol is again divided in the same way and the second digit is determined as

Table 8.2. *A message source produces the symbols a, b, c, and d with given probabilities. Last row: the Fano binary code that represents these symbols.*

Symbol	a	b	c	d
Probabilities p_i	0.5	0.2	0.2	0.1
$\log_2(1/p_i)$	1	2.32	2.32	3.32
Binary representation	0	10	110	111

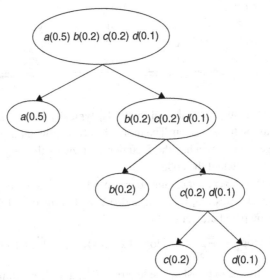

Figure 8.2. A tree graph that represents the Fano algorithm for binary coding of symbols, *a, b, c,* and *d,* that appear with probabilities 0.5, 0.2, 0.2, and 0.1.

before. The divisions continue until each symbol is on a separate leaf. In this way we find the binary representations listed in the last row of Table 8.2.

Fano code is, by design, uniquely decipherable. For example, the sequence *111010100111000* can only be deciphered as *dabbadaaa* when read from left to right using the Fano code of Table 8.2. Some apparently more efficient ways of coding symbols into binary numbers are not decipherable. Consider, for instance, using a translation rule, $a \leftrightarrow 0$, $b \leftrightarrow 1$, $c \leftrightarrow 10$, and $d \leftrightarrow 11$, shorter than that provided by Fano code, to decipher *111010100111000*. Because we do not know whether the leading *1* stands by itself or is the first digit of *11*, we do not know whether the first letter is *b* or *d*.

Text written in the Fano code of Table 8.2 has an average of

$$0.5 \times 1 + 0.2 \times 2 + 0.2 \times 3 + 0.1 \times 3 = 1.80 \text{ bits} \qquad (8.17)$$

of Shannon entropy per transmissible symbol – a little more than the 1.76 bits of Shannon entropy per symbol produced by the original source. More efficient coding schemes exist, but they cannot compress the Shannon entropy of the transmissible code below the Shannon entropy of the source.

Comparing the structure of the Shannon entropy per binary digit in Eq. (8.16) with the structure of the average entropy per transmissible binary digit in Eq. (8.17) illustrates why Fano code aims to represent the ith symbol of the original alphabet with a coded symbol of $\log_2 (1/p_i)$ binary digits. If this were exactly possible the average Shannon entropy per character of the binary code would exactly match the Shannon entropy per character of the original alphabet.

8.6 Data compression and error correction

Suppose you periodically receive emails from a verbose and ne'er-do-well but beloved friend. After a while, you realize that the content of your friend's emails always amounts to one of three messages that appear with equal frequency: (1) "Send money." (2) "Can you put me up tomorrow?" and (3) "Cheers!"

One day you receive an email from an unrecognizable source and because you are a technophile you know ahead of reading the message that it contains 1000 characters each one of which is coded with a standard byte, that is, with 8 bits. At this point the Shannon entropy of the source appears to be $(1000 \text{ characters}) \times (8 \text{ bits/character})$ or 8000 bits. Then you look more closely at the message. It's from your ne'er-do-well friend and, given its source, you realize, regardless of the number of its characters, the message will be one of the three standard ones. The entropy of the message has suddenly collapsed to $\log_2 3$ or 1.58 bits – bits you can recover quickly by scanning the message. In effect, you have, by recognizing the redundancy of much of its text, compressed the entropy of the message by a factor of 8000/1.58 or 5000. Of course, in compressing the email some real, if relatively unimportant, information contained in the details of your friend's message has been lost. Yours is a *lossy compression*.

Lossy data compression

Lossy data compression algorithms are especially useful in reducing audio, video, and multimedia files that typically store multiple gigabytes of information. Audio compression algorithms, for instance, eliminate acoustic frequencies to which the human ear is partially or completely insensitive. Video compression algorithms eliminate bits that code parts of images that are so redundant their content can be accurately guessed. Typical lossy compression ratios of the Shannon entropy of multimedia files range from 4 to 8.

Error correction

However, redundant information is not all bad. Redundancy is, after all, one protection against the errors that accumulate in a noisy communication channel. The best general error correction strategy is to eliminate the unimportant and repeat the important, that is, select compression followed by select decompression. Long before the age of digital computers Professor Will Strunk, author of the first draft of the bestselling text *Elements of Style*, had grasped this essential point. According to E. B. White, Professor Strunk's favorite rule of style was: "Omit needless words." Strunk

> ... omitted so many needless words, and omitted them so forcibly and with such eagerness and obvious relish, he often seemed in the position of having shortchanged himself – a man with nothing left to say yet with time to fill, a radio prophet who had outdistanced the clock. Will Strunk got out of this predicament by a simple trick: he uttered every sentence three times.... "Rule Seventeen: Omit needless words! Omit needless words! Omit needless words!"

Professor Strunk's habit of repeating each message twice works well with binary code. Consider, for instance, that noise may randomly switch a binary digit, 1 or 0, from one to the other during transmission. If the error rate is such that no more than one symbol in three is switched, sending each digit three times will always allow the receiver to recover the original message. For if the original digit is 1, repeating it twice produces 111. If no more than one of these digits is received incorrectly, the message received will be 111 (no error), or 011, 101, or 110 (one error). In any case the correct digit, 1, can always be recovered since it is the digit appearing most often.

Repeating a binary digit twice in order to correct transmission errors increases the entropy of the received message source, relative to the entropy of the original message source, by a factor of 3. If the error rate is even smaller than one in three, we can improve on this ratio.

Suppose, for instance, that noise changes no more than one binary digit in five. To minimize the required redundancy while still allowing the transmitted code to be corrected we send binary data in pairs, say, X_1, X_2 where X_1 and X_2 are the intended binary digits 0 or 1. Sending a pair twice produces X_1, X_2, X_1', X_2' where, by design, $X_1' = X_1$ and $X_2' = X_2$. Suppose then that the received sequence X_1, X_2, X_1', X_2' is such that $X_1' = X_1$ and $X_2' \neq X_2$ – the latter because of an error. Then the receiver would know only that either the second binary digit X_2 or the fourth binary digit X_2' had been switched during transmission. We need to add one more binary digit to the sequence in order to resolve this ambiguity. In place of X_1, X_2, X_1', X_2' the coder sends the five-digit sequence $X_1, X_2, X_1', X_2', X_3$ where by design $X_2' = X_2$, $X_2' = X_2$, and X_3 is chosen so that

$$X_1 + X_2 + X_3 = even,$$
(8.18)

where *even* stands for an even number, either 0 or 2, depending upon the values of X_1 and X_2 at the source. Then if $X_2 \neq X_2'$ and $X_1 + X_2 + X_3 = even$ the receiver knows that X_2 is correct and X_2' incorrect. If, on the other hand $X_2 \neq X_2'$ and $X_1 + X_2 + X_3 = odd$, where *odd* stands for an odd number (1 or 3), the receiver knows that the received value of X_2 is incorrect and X_2' is correct. Therefore, sending a pair of binary digits in five-digit sequences increases the entropy of the received code by at most a factor of 5/2 – an improvement over the factor of 3 required by the tactic of repeating twice. Smaller error-generation rates allow the coder to reduce this ratio further.

8.7 Missing information and statistical physics

Information theory provides us with a language parallel in structure and vocabulary to the language with which we have described physical systems in Chapters 2–7. We explored part of this parallelism in Section 8.3. Here we continue this exploration.

A message source is a set of symbols or messages the ith one of which appears in the set with a frequency that reflects its probability p_j. We can also think of a physical system, more exactly the ensemble of a physical system, as a message source. Imagine that we could question a physical system and at each query be provided with a message that describes the system microstate – again the ith message describing the ith microstate appearing with probability p_j. For this reason we can speak either of the entropy S or, equivalently, of the missing information,

$$H = \frac{S}{k}$$
$$= -\sum_j p_j \ln p_j,$$
(8.19)

of a source that produces messages that describe the ensemble of a physical system. The missing information described by (8.19) is measured in *nats*. However, the parallelism between information theory and statistical physics goes beyond replacing one name with another. Information theory gives us a new way of thinking about the statistical mechanics of physical systems.

The main task of the information theoretic approach to statistical physics is to assign the probabilities p_j that characterize the accessible microstates of a physical system. In making this assignment the information theorist is guided by several principles that, in effect, replace the guidance given the physicist by the first and second laws of thermodynamics. These principles are that the probabilities p_j (1) quantify our knowledge of the ensemble, and (2) do so without introducing unwarranted assumptions or biases. Of course, (3) the probabilities must also be normalized so that

$$\sum_j p_j = 1, \tag{8.20}$$

where the sum is over all microstates consistent with a given macrostate or ensemble.

In general we avoid unwarranted assumptions or biases by making the probabilities p_j as nearly equal to each other as the constraints permit. To do so we maximize the missing information H subject to (8.20) and any other constraints that quantify our knowledge of the ensemble. Recall that *missing information*, *uncertainty*, and *entropy* are synonyms. Therefore, maximizing the missing information is equivalent to acknowledging all of our uncertainty. The principle of acknowledging all of our uncertainty by maximizing the missing information is often called the *principle of maximum entropy*.

Microcanonical ensemble

As a simple first example, consider the case in which we know nothing about a physical system except that it occupies a set of microstates. We seek an unbiased assignment of the probabilities p_j with which the system realizes these microstates. Because we want to avoid introducing bias, we maximize the system's missing information H subject only to the normalization condition (8.20). Thus, we maximize the constrained missing information

$$-\sum_j p_j \ln p_j + \alpha \left[1 - \sum_j p_j \right] \tag{8.21}$$

by making its derivatives with respect to p_i vanish, and find that

$$-\ln p_i - 1 - \alpha = 0 \tag{8.22}$$

and, therefore, that the microstate probabilities

$$p_i = e^{-(1+\alpha)} \tag{8.23}$$

are equiprobable. Since the second derivatives of (8.21) with respect to the probabilities p_i are negative, (8.23) identifies a relative maximum of (8.21) and not merely a stationary value. The result (8.23) and the normalization condition (8.20) produce $p_1 = p_2 = \cdots = p_\Omega = 1/\Omega$. Therefore,

$$
\begin{aligned}
H &= -\sum_j p_j \ln p_j \\
&= \sum_{j=1}^{\Omega} \left(\frac{1}{\Omega}\right) \ln \Omega \\
&= \ln \Omega,
\end{aligned}
\tag{8.24}
$$

where Ω is the number of microstates accessible to the system.

Canonical ensemble

The energy of a system in thermal equilibrium with its environment may vary and in doing so realize different microstates j with probability p_j and energy E_j. In general, these microstates are occupied in rapid succession. Since the average energy $\langle E \rangle \left[= \sum_j p_j E_j \right]$ of the system is what we mean by the system's internal energy, we adopt

$$
\langle E \rangle = \sum_j p_j E_j
\tag{8.25}
$$

and the normalization condition (8.20) as constraints that define the canonical ensemble.

We again seek an unbiased assignment of the probabilities p_i that characterize this system. Thus we maximize the missing information of this ensemble subject to the two constraints (8.20) and (8.25). Because we have two constraints we employ two Lagrange multipliers: α and β. Maximizing the constrained missing information

$$
-\sum_j p_j \ln p_j + \alpha \left[1 - \sum_j p_j \right] + \beta \left[E - \sum_j p_j E_j \right]
\tag{8.26}
$$

means following the same procedure we have already followed twice, in Section 3.4 and in Section 5.4. The result is a probability distribution

$$
p_i = \frac{e^{-E_i/kT}}{\sum_j e^{-E_j/kT}}
\tag{8.27}
$$

that recovers the probability p_i that a system in a canonical ensemble will realize microstate i with energy E_i.

The information theoretic approach to statistical mechanics

The information theoretic approach to statistical mechanics reproduces these and other standard results. But does it produce any new results? The language of information theory seems to cast the statistical mechanics of physical systems into a more general framework. Certainly E. T. Jaynes makes a strong case that statistical mechanics is primarily an example of minimizing bias. The chief tool of this minimization is what he calls the principle of maximum entropy. (For details see his paper "Information theory and statistical mechanics" listed in Appendix VI.) But questions remain and whether or not, and if so in what ways, message sources and ensembles representing physical systems are connected in more substantial ways remains a matter of investigation.

Problems

8.1 Deriving the logarithm

Show that the only continuous functions $f(x)$ and $g(x)$ on the domain $x \geq 0$ that satisfy the relation (8.3), that is, $n\,f(s) = g(s^n)$, are $f(x) = g(x) = c \ln x$, where c is an arbitrary constant. (Suggested strategy: Take the partial derivatives with respect to n and with respect to s of the equation $n\,f(s) = g(s^n)$, eliminate the function $g(s^n)$ from these two equations, and solve for $f(s)$.)

8.2 Information content

(a) A message source is composed of the following equiprobable messages: 11, 12, 13, 14, 21, 22, 23, 24, 31, 32, 33, 34, 41, 42, 43, and 44. What is the information content of this message source in bits? (b) A message source is composed of the following equiprobable messages: 11, 12, 13, 14, 21, 22, 23, and 24. What is the information content of this message source in bits?

8.3 Block coding

(a) What is the information content in bits of a message source composed of all possible message sequences four characters long in which each position in the sequence can be filled with one of 27 equally probable symbols (the 26 letters of the alphabet and one space)? (b) Suppose the message source described in part (a) is translated into another message source composed of all possible message sequences 12 characters long in which each position in the sequence can be filled with one of three equally probable symbols (the numerals 1, 2, and 3).

Table 8.3. *A rule that translates one code into another. Used in Problem 8.3.*

$a = 111$	$b = 112$	$c = 113$	$d = 121$	$e = 122$	$f = 123$	$g = 131$	$h = 132$	$i = 133$
$j = 211$	$k = 212$	$l = 213$	$m = 221$	$n = 222$	$o = 223$	$p = 231$	$q = 232$	$r = 233$
$s = 311$	$t = 312$	$u = 312$	$v = 313$	$w = 321$	$x = 322$	$y = 331$	$z = 332$	$space = 333$

(See Table 8.3 for the translation rule.) What is the information content in bits of this message source? (c) Show that the information content of these two message sources are equal.

8.4 Information from a grade?

What is the Shannon entropy in bits conveyed to a student who receives a grade (a) on the scale A, B, C, D, or F and (b) on the scale A, A−, B+, B, B−, C+, C, C−, D+, D, D−, and F. Assume each grade is equally probable.

8.5 Impossible outcome

Suppose one message contained in a source is impossible, that is, appears with zero probability. How much does the impossible message contribute to the entropy of the source? (Hint: Start with the relation

$$H\left(p_1, p_2, \ldots, p_s, 0\right) = H\left(p_1, p_2, \ldots, p_s\right) - \lim_{p \to 0} p \ln p$$

and use L'Hopital's rule to evaluate the limit.)

8.6 Efficient Fano code

Suppose a message source produces the letters a, b, c, and d with, respectively, the following probabilities: 1/2, 1/4, 1/8, and 1/8. (a) Find the Shannon entropy of this source. (b) Find the Fano binary code that represents these letters. (c) Find the average entropy per symbol of the Fano code.

8.7 Hamming error correction

When the error production rate of a noisy channel is no more than one binary digit in each sequence of seven, it is possible to send four correctable binary digits of data in each sequence of seven binary digits: X_1, X_2, X_3, X_4, X_5, X_6, X_7. Digits X_1, X_2, X_3, and X_4 are data and the remaining three binary digits X_5, X_6, and X_6 ensure correctability. These 7-digit binary sequences are called Hamming (7,4) code after the computer scientist Richard Hamming (1915–1998) who invented this correction scheme in 1950. The correction digits X_5, X_6, and X_6 are determined by the requirements

$$X_1 + X_2 + X_3 + X_5 = even,$$
$$X_1 + \_\_ + \_\_ + X_6 = even,$$

and

$$X_1 + \_\_ + \_\_ + X_7 = even,$$

where *even* stands for an even number and each blank (__) stands for either X_1, X_2, X_3, or X_4. (a) Fill in the blanks in the above equations with a set of binary digits that make the data correctable. (b) Assuming that the binary digits 0 and 1 appear with equal probability, find the Shannon entropy of a 7-digit sequence of Hamming code and the Shannon entropy of the 4-digit sequence of binary data it represents.

Epilogue
What is entropy?

While it remains true that "entropy is not a localized, microscale phenomenon at which we can point, even in our imaginations, and say, 'Look! There is entropy'" and that, "if we insist on trying to understand a subject in ways inconsistent with its nature, we will be disappointed," the eight chapters of this guide have prepared us to give a constructive answer to the question "What is entropy?"

Any short description of entropy will necessarily be figurative. After all, one task of a figure of speech is to transfer a complex meaning from an extended description to a word or short phrase. In fact, we have already considered several figurative descriptions of entropy that are appropriate in special contexts: *transformation content, disorder, uncertainty, spread in phase space,* and *missing information. Transformation content* was Clausius's way of referring to how the entropy function indicates the direction in which an isolated system may evolve. *Spread in phase space*, while appropriate for statistical systems, depends upon familiarity with the technical concept of phase space.

Disorder has long been a popular synonym for entropy. But recently *order* and *disorder* as describing low and high entropy systems have fallen into disfavor. This is because scientists have become fascinated with isolated systems that generate apparent order from apparent disorder. For instance, consider a thoroughly shaken bottle of water and olive oil. When left undisturbed, the water and olive oil begin to separate into distinct layers with the less dense olive oil on top. Yet even in this process the entropy of the oil–water system increases. Thus, while *order* and *disorder* are suggestive, they can mislead.

Missing information is a measure of the information required to choose one particular microstate from an ensemble of possible microstates or one message from a source of possible messages. For this reason missing information is a powerful metaphor that draws our attention to the many parallels between the concepts of statistical physics and information theory. The concept of

159

uncertainty plays a similar role. The larger and more varied the ensemble from which a system can be chosen and the larger and more varied the message source from which a message can be drawn, the more uncertain the outcome and the higher the entropy.

Entropy as *possibility* is my favorite short description of entropy because possibility is an apt word and, unlike *uncertainty* and *missing information*, has positive connotation. Thus, according to the second law of thermodynamics, an isolated thermodynamic system always evolves in the direction of opening up new possibilities. And the larger the set of possibilities from which either a microstate or a message can be realized or chosen, the larger the entropy of the physical system or the Shannon entropy of the message source.

A formal definition of *entropy* in terms of possibility is: *Entropy is an additive measure of the number of possibilities available to a system.* Thus the entropy of a physical system is an additive measure of the number of possible microstates that can be realized by that system. And the entropy of a message source is an additive measure of the number of possible messages that can be chosen from that message source. The image on the cover of this book almost perfectly captures the sense of entropy as possibility. As the constraints that inform a living organism dissolve, the entropy of the organism increases. As the flower dies, its seeds are scattered in the breeze. Yet even in this death, new possibilities are sown.

Appendix I

Physical constants and standard definitions

Physical constants in SI units

Name	Symbol and value
Boltzmann constant	$k = 1.38 \times 10^{-23}$ m^2 kg/(s^2 K)
Planck constant	$h = 6.63 \times 10^{-34}$ m^2 kg/s
Gas constant	$R = 8.31$ J/(K mole)
Electron mass	$m_e = 9.11 \times 10^{-31}$ kg
Electron charge	$e = 1.60 \times 10^{-19}$ C
Standard acceleration of gravity	$g = 9.81$ m/s^2
Speed of light in vacuum	$c = 3.00 \times 10^8$ m/s
Radiation constant	$a = 7.57 \times 10^{-16}$ kg/(m s^2 K^4)
Atomic mass unit	u $= 1.66 \times 10^{-27}$ kg

Standard definitions

Name	Symbol and definition
Mechanical equivalent of heat	$J = 4.19$ J/cal
Atmospheric pressure	1 atm 1.01 $\times$ 10^5 Pa
	$= 760$ torr
	$= 760$ mm Hg
	$= 14.7$ lbs/in^2
Bar	1 bar $= 10^5$ Pa
Electronvolt	1 eV $= 1.60 \times 10^{-19}$ J

Appendix II
Formulary

$$\ln(xy) = \ln x + \ln y$$

$$\ln\left(\frac{x}{y}\right) = \ln x - \ln y$$

$$\log_b y = \frac{\ln y}{\ln b}$$

$$\int_0^\infty \frac{x^3 dx}{e^x - 1} = \frac{\pi^4}{15}$$

$$\int_0^\infty \frac{x^2 dx}{e^x - 1} = 2Z(3) \approx 2.404$$

$$\int_0^\infty x^2 \ln(1 - e^{-x}) dx = \frac{-\pi^4}{45}$$

$$\int_{-\infty}^\infty e^{-x^2} dx = \sqrt{\pi}$$

$$e^x = 1 + x + \frac{x^2}{2!} + \frac{x^3}{3!} + \cdots \quad \text{for} \quad x < 1$$

$$\ln(1 + x) = x - \frac{x^2}{2} + \frac{x^3}{3} - \frac{x^4}{4} + \cdots \quad \text{for} \quad x < 1$$

Appendix III
Glossary

Absolute entropy: The unique value of entropy that describes the macrostate of a system. Same as *Third law entropy.*

Absolute temperature: A temperature determined by the efficiency of a Carnot engine operating between a heat reservoir in a standard state and a reservoir in thermal equilibrium with the object whose temperature is to be determined. Same as *Thermodynamic temperature.*

Additivity: A thermodynamic variable is additive when its value is the sum of the corresponding values describing each part of a composite system. The entropy S and energy E of a thermodynamic system are additive. The component parts to which additivity applies are sometimes but not always spatially separate. Related to *Extensive variable.*

Average energy approximation: A description of the ideal quantum gases in terms of the number of microstates $V(E/N)^{3/2}(4\pi em/3h^2)^{3/2}$ available to a single particle that occupies volume V and possesses the average energy E/N of a system of N particles. Here e stands for the base of the natural logarithm, m for the mass of the particle, and h for Planck's constant.

Bandgap energy: The separation in energy between the valence and conduction band electrons in a semiconductor.

Blackbody radiation: Electromagnetic radiation in equilibrium with a material system at thermodynamic temperature T. The thermodynamic variables of a system of blackbody radiation are its energy E and its volume V. Also known as equilibrium radiation.

Boltzmann distribution: The distribution of microstate probabilities $p_i = e^{-\beta E_i}/\sum_i e^{-\beta E_i}$ for microstate energy E_i and where $\beta = 1/kT$ appropriate for a system that is in thermal equilibrium with its environment at temperature T. When the system is a single particle of an ideal gas the Boltzmann distribution is a Maxwell–Boltzmann distribution. Same as *Canonical distribution.*

Boltzmann entropy: The third law compliant entropy of the macrostate of a system described by $S = k\ln\Omega$ where Ω is the number of equally probable microstates of the macrostate of an isolated system.

Bose–Einstein condensate: A low temperature phase of matter composed of *bosons*, that is, particles that obey Bose–Einstein statistics. At sufficiently low temperatures, bosonic matter condenses into a macrostate in which all its particles occupy one single-particle microstate.

Bosons: Particles that obey Bose–Einstein statistics. According to Bose–Einstein statistics, any number of identical bosons can occupy the same single-particle microstate.

Canonical distribution: Same as *Boltzmann distribution.*

Canonical ensemble: A collection of identical systems that individually realize their microstates with probabilities $p_i \left[\propto e^{-E_i/kT} \right]$ characteristic of a system in equilibrium with an environment at temperature T.

Carnot efficiency: The maximum efficiency of a heat engine that operates between two heat reservoirs. The Carnot efficiency is given by $\varepsilon_{\text{Carnot}} = 1 - T_C/T_H$ where T_C is the temperature of the colder temperature reservoir and T_H is the temperature of the hotter reservoir.

Carnot's theorem: Asserts that the most efficient heat engine operating between a pair of heat reservoirs is one that operates reversibly.

Chemical potential: The intensive variable brought into existence by the variability of the number of particles that compose a system. The chemical potential is the increment in energy when a fluid system grows by one particle given that its entropy and volume remain constant. Symbolically, the chemical potential μ is described by $\mu = \left(\partial E/\partial N \right)_{S,V}$ or by $\mu = -T \left(\partial S/\partial N \right)_{V,E}$.

Classical microstate: The description of a system that consists of identifying the location of the system's distinguishable particles within arbitrarily constructed phase space cells.

Classically distinguishable particles: Particles that can, in principle, always be distinguished from one another because of their distinct trajectories in space and time.

Complementarity: The principle that quite different, even contradictory, physical models can describe a single phenomenon. The wave and photon pictures describe electromagnetic radiation in complementary fashion.

Compton effect: The scattering of an X-ray photon from a free or essentially free electron. In the process the X-ray photon transfers some momentum and energy to the electron.

Correspondence principle: Quantum expressions reduce to classical ones in an appropriate limit.

Degenerate, Degeneracy: Words that literally mean "not normal." In statistical mechanics these words refer to system or single-particle microstates having the same energy. Thus, two particles are degenerate when they occupy the same single-particle microstate or different single-particle microstates with the same energy. Less frequently, the multiplicity Ω of an isolated system is called its degeneracy since all the system's microstates have the same energy. Also, a ground macrostate, that is, a $T = 0$ macrostate, is degenerate when it is composed of more than one microstate.

Efficiency: The efficiency of a heat engine operating between two heat reservoirs is the ratio of the work performed by the engine in one cycle W to the heat extracted from the hotter reservoir in one cycle Q_H, that is, W/Q_H.

Einstein solid: A crystalline array of atoms or molecules each one of which oscillates simple harmonically and independently in three dimensions with a common frequency. The energy of an oscillator in an Einstein solid is quantized.

Einstein temperature: The parameter $h\nu_o/k$ where the symbol ν_o is the common frequency with which all the atoms of an Einstein solid vibrate. The Einstein temperature was used by Einstein to fit the specific heat of an Einstein solid to available data.

Ensemble: A large number of fictional copies of a system. The frequency with which the copies in an ensemble are found in a particular microstate reflect the probability of that microstate. Some standard ensembles have been given special names. The *microcanonical* ensemble, for instance, contains copies of a completely isolated system with a given energy, and the *canonical* ensemble contains copies of a system in thermal equilibrium with an environment at a given temperature T.

Entropy: A measure of the *irreversibility* of the thermodynamic evolution of an isolated system. Also, a measure of the *spread in phase space*, the *uncertainty*, or the *missing information* of the macrostate of a system. An additive measure of the number of configurations available to a system, or, equivalently, an additive measure of the number of possibilities open to a system.

Equation of state: A relation among the thermodynamic variables that describe a system macrostate.

Equilibrium state: A time-independent macrostate of a thermodynamic system. Same as *Thermodynamic state*.

Equipartition theorem: Asserts that every term in a particle's energy that is quadratic in a phase space coordinate adds $kT/2$ to the system's internal energy. The equipartition theorem is observed in classical systems when the phase space integrations of the partition function can be extended indefinitely.

Extensive variable: A thermodynamic variable is extensive when its value is the sum of the values describing each part of a homogeneous composite system. Extensivity is a special case of *Additivity*. For example, the energy E, volume V, particle number N, and entropy S of a simple fluid are extensive variables.

Fano code: A simple method of translating an alphabet of symbols into binary digits that aims to keep the entropy of the binary digits as close as possible to the Shannon entropy of the original alphabet.

First law of thermodynamics: The law of conservation of energy applied to a thermodynamic system.

Fluid: Any thermodynamic system that can be described with only two state variables: an isotropic pressure P and volume V. Same as *Simple fluid*.

Fundamental constraint (of a fluid system): The relation $dE = T\,dS - P\,dV$ among the differentials of the thermodynamic variables, energy E, temperature T, entropy S, pressure P, and volume V of a fluid system. When a fluid system is not closed, its fundamental constraint becomes $dE = T\,dS - P\,dV + \mu\,dN$ where μ is the chemical potential and N is the number of particles composing the system.

Fundamental postulate (of statistical mechanics): Asserts that all microstates composing the macrostate of an isolated system are equally probable.

Gibbs entropy formula: The relation $S = -k\sum_j p_j \ln p_j$ that determines the entropy S from the probabilities p_j $[j = 1, 2, \ldots]$ that a system will occupy its jth microstate. The entropy determined from the Gibbs entropy formula is third law compliant.

Ground state: The macrostate of a system in the $T \to 0$ limit, that is, with the lowest possible energy.

Ground state pressure: The pressure that persists in the $T \to 0$ limit of an ideal Fermi gas. Sometimes called the *degeneracy pressure*.

Ground state energy: The energy that persists in the $T \to 0$ limit of an ideal Fermi gas. Sometimes called the degeneracy energy.

Heat: Energy transferred by heating or cooling.

Heat engine: A device that uses a temperature difference to produce work.

Heat reservoir: Same as *Temperature reservoir*.

Hole: The vacancy left in the valence band of a semiconductor when an electron is excited from the valence band into the conduction band. Both valence band holes and conduction band electrons contribute to the conductivity of a semiconductor.

Information: Ralph Hartley's word for the information content of a message source composed of equally probable symbols or messages. Information is the logarithm of the number of equally probable symbols or messages in a message source.

Intensive variable: A thermodynamic variable is intensive when its value remains constant as the system is scaled up or down. Temperature T, pressure P, and chemical potential μ are intensive fluid variables.

Internal energy: The energy E contained in a thermodynamic system. The internal energy excludes the energy associated with the motion or the position of the system as a whole.

Intrinsic semiconductor: A crystalline semiconductor with no defects or impurities.

Irreversible process: A thermodynamic process whose direction can be reversed only by making finite changes in the state of the system or its environment. A process that is not quasistatic or one that proceeds with friction, dissipation, or hysteresis is irreversible. See also *Reversible process*.

Loschmidt's paradox: All thermodynamic systems are composed of fundamental particles, all thermodynamic processes are composed of the interactions of fundamental particles, and all interactions of fundamental particles are reversible; but all non-idealized thermodynamic processes are irreversible.

Lossy compression: The reduction of the Shannon entropy of a message source by eliminating redundant or inessential information.

Macrostate: A set of microstates that can be described with a small set of thermodynamic variables.

Maxwell's demon: A small being, or inanimate automaton, that can decrease the entropy of a system by exploiting the fluctuations in its parts.

Message source: The set of all possible messages from which particular messages can be drawn. An ensemble of messages.

Microcanonical ensemble: An ensemble composed of copies of a completely isolated system with a given energy E.

Microstate: The most precise possible description of the arrangement of the atoms and molecules of a system. See also *Classical microstate*.

Missing information: The information needed to specify the microstate of a system given its macrostate or the information needed to select a particular message from a message source. Same as *Shannon entropy*. Also the third law entropy of a physical system divided by Boltzmann's constant k.

Multiplicity: The number of microstates that an isolated system can occupy. Sometimes called the *Degeneracy* or the *Thermodynamic probability* or the *Statistical weight* of a system. While the traditional symbol for multiplicity is Ω, the symbol for multiplicity that appears on Boltzmann's tombstone is W.

Occupancy: The number of particles in a multi-particle system divided by the number of accessible single-particle microstates. The average occupation number of a system.

Occupation number (of a cell): The number of particles of a multi-particle system that occupy the same cell in a 6-dimensional phase space common to all the particles of the system.

Partition function: The quantity $Z \left[= \sum_i e^{-E_i/kT} \right]$ where the sum is over all the microstates of the system and E_i is the energy of the system when in microstate i. Compare with *Single-particle partition function*.

Pauli principle: Asserts that no two atomic electrons can occupy the same single-particle microstate.

Phase space: A multi-dimensional space whose coordinates are the position and momenta of the particles composing the system.

Phonons: Bundles of energy whose size is determined by the units of energy that can be absorbed or rejected by the oscillators that compose a solid.

Photoelectric effect: High frequency light shines on a metallic surface and ejects electrons.

Photon: A quantum of light. The energy ε and frequency v of a photon are related by $\varepsilon = hv$ while its momentum p and frequency v are related by $p = hv/c$.

Quantum indistinguishability: Two particles are quantum mechanically indistinguishable when exchanging them would not lead to experimentally observable results. If two particles are quantum mechanically indistinguishable, they are indistinguishable for the purpose of counting microstates.

Quasistatic process: An indefinitely slow process. A system experiencing a quasistatic process passes through a continuum of thermodynamic states.

Radiation constant: The proportionality constant between the energy density E/V of blackbody radiation and the fourth power of its absolute temperature T^4. The radiation constant is usually denoted with the symbol a. Thus, $E/V = aT^4$. Quantum statistical arguments show that $a = 8\pi^5 k^4 / (15c^3 h^3)$.

Random variable: A variable that realizes a range of values according to a set of probabilities.

Reversibility paradox: Same as *Loschmidt's paradox*.

Reversible process: A thermodynamic process whose direction can be reversed by making only an indefinitely small change in the system or in its environment. A reversible process necessarily proceeds indefinitely slowly, that is, quasistatically, and without friction, dissipation, or hysteresis. See *Irreversible process*.

Semi-classical limit: The limit of a quantum statistical expression in which kT is much greater than a single quantum of energy.

Shannon entropy: A measure of the information content of a message source. The Shannon entropy of a source of one-character messages is given by $H = -c \sum_i p_i \ln p_i$ where p_i is the probability that the character is realized by the ith symbol and c is a constant that determines the unit of information.

Simple fluid: Same as *Fluid.*

Single-particle partition function: The quantity $Z_1 \left[= \sum_j e^{-\varepsilon_j/kT} \right]$, where the sum is over all the microstates of a single-particle in a phase space common to all particles and where ε_j is the energy of the single particle in microstate j. Single-particle partition functions are useful whenever the particles of a system occupy their positions in a common phase space independently of each other except for system-wide constraints such as those imposed on the total energy E and particle number N.

Spin-statistics theorem: Asserts that particles with even units of spin, called bosons, can occupy the same single-particle microstate and particles with odd units of spin, called fermions, cannot occupy the same single-particle microstate. Wolfgang Pauli (1900–1958) discovered the spin-statistics theorem in 1940.

Statistical weight: See *Multiplicity.*

Stigler's law of eponymy: Asserts that no discovery is named after its original discoverer.

Sure variable: A variable that realizes a single, definite value. Not a *Random variable.*

Temperature, empirical: Hotness measured on a scale defined by the behavior of a physical system, that is, as determined by a *Thermometer.*

Temperature reservoir: A system that maintains the same temperature regardless of how much heat the system absorbs or rejects. A temperature reservoir has indefinitely large heat capacity. Same as *Heat reservoir.*

Thermodynamic probability: Max Planck's term for *Multiplicity.* Other synonyms are thermodynamic weight and statistical weight. Also equivalent to the *Degeneracy* of an isolated system.

Thermodynamic system: A macroscopic system that observes the fundamental postulate of statistical mechanics. A thermodynamic system can be described with a small set of thermodynamic variables.

Thermodynamic state: Same as *Equilibrium state.*

Thermodynamic temperature: Same as *Absolute temperature.*

Thermodynamic weight: Same as *Multiplicity.*

Thermometer: A physical system that associates the size of a convenient thermodynamic variable, such as volume, resistance, or color, with a unique number. The way in which the thermodynamic variable is associated with a unique number defines a temperature scale.

Third law entropy: A formulation of the entropy that exploits our freedom to adopt a conventional value for the entropy in the limit of vanishing thermodynamic temperature. The formulations $S = k \ln \Omega$ and $S = -\sum_i p_i \ln p_i$ are third law entropies. Same as *Absolute entropy.*

Appendix IV

Time line

1742	Anders Celsius (1701–1744) devises an empirical temperature scale that assigns 0 to the boiling point of water and 100 to its freezing point.
1745	Carolus Linnaeus (1701–1778) turns Celsius's scale upside down.
1803	John Dalton (1766–1844) enunciates a "law of simple multiple proportions" that summarizes the way different elements combine and suggests the existence of atoms and molecules.
1819	The law of Dulong and Petit is discovered. According to this law the molar specific heat capacity C_V/n of a solid is a universal constant $3R$ $\left[\approx 25 \text{ J}/(\text{K mole})\right]$ where R is the so-called gas constant.
1824	Sadi Carnot (1796–1832) publishes *On the Motive Power of Fire* in which he articulates a version of the second law and proves Carnot's theorem.
1840–1850	James Joule's precision experiments compel acceptance of the first law of thermodynamics.
1848	William Thomson (1824–1907) (later known as Lord Kelvin) invents the concept of absolute or thermodynamic temperature.
1850	Rudolph Clausius (1822–1888) articulates a version of the second law and distinguishes between the first and second laws of thermodynamics.
1851	William Thomson (Lord Kelvin) articulates a version of the second law.
1865	Rudolph Clausius discovers the concept of entropy.
1871	James Clerk Maxwell (1831–1879) publicly introduces Maxwell's demon in his book *Theory of Heat*. Maxwell's purpose is to demonstrate that entropy is a statistical or probabilistic quantity.
1876	Johann Loschmidt (1821–1895) asks, "Why, if the interactions of elementary particles are reversible, are thermodynamic processes irreversible?" This disjunct between microscopic and macroscopic physics is known as Loschmidt's paradox.

1877   Ludwig Boltzmann (1844–1906) publishes a paper *On the relation of the second law of the mechanical theory of heat and the probability calculus with respect to the theorems on thermal equilibrium* in which he first introduces the relation of entropy to macrostate multiplicity and identifies the most probable occupation number macrostate with the state of thermal equilibrium.

1900   Max Planck (1858–1947) announces a derivation of the spectral energy density of blackbody radiation in which he is forced to assume that light is emitted from matter in discrete, quantized amounts.

1902   J. Willard Gibbs (1839–1903) publishes *Elementary Principles of Statistical Mechanics*. This book presents the major methods and applications of classical statistical mechanics and introduces the concept of an ensemble.

1905   Albert Einstein (1879–1955) publishes groundbreaking papers on Brownian motion, special relativity, and the photoelectric effect.

1906   Walther Nernst (1864–1941) formulates what has since became known as the third law of thermodynamics.

1907   Einstein proposes a simple quantum model of a crystalline solid, the so-called Einstein solid, which initiates the modern study of condensed phase physics.

1912   The Sackur–Tetrode equation for the extensive entropy of an ideal gas with quantized phase space is introduced.

1923   Arthur Holly Compton (1892–1962) observes the Compton effect in which a photon striking a free electron transfers energy and momentum to the electron.

1924   July. S. N. Bose (1894–1974) derives the spectral energy density of blackbody radiation by counting the microstates of a gas of photons.

1924   July. Einstein applies Bose's method of counting microstates to a gas of particles with non-zero rest mass.

1925   January. Wolfgang Pauli (1900–1958) proposes the Pauli principle according to which no two electrons within an atom can share the same single-particle microstate.

1925   January. Albert Einstein predicts the existence of a new phase of matter later called the Bose–Einstein condensate.

1925   July. Werner Heisenberg (1901–1976) formulates the first version of the new quantum mechanics.

1926   January. Erwin Schrodinger (1887–1961) introduces the wave mechanics version of quantum mechanics.

1926   February. Enrico Fermi (1901–1954) describes the statistics of particles no two of which can share the same single-particle microstate, particles later known as fermions.

1926   August. P. A. M. Dirac (1902–1984) relates symmetric and antisymmetric wave functions with, respectively, Bose–Einstein and Fermi–Dirac statistics.

1928   R. V. L. Hartley (1988–1970) defines a measure $H \left[\propto \ln s^n \right]$ of the information content of a message source that produces messages of length n each character of which can realize s equally probable symbols.

1929   Leo Szilard (1898–1964) analyzes the interaction of Maxwell's demon with a one-particle gas. Szilard's paper initiates a line of research, active to this day, on the connections between information and entropy.

1933   Ludwig Boltzmann's tombstone in the Vienna Central Cemetery, inscribed with the equation $S = k \log W$, is dedicated.

1940   Wolfgang Pauli discovers the spin-statistics theorem according to which particles with even units of intrinsic spin (bosons) can share the same single-particle microstate while particles with odd units of intrinsic spin (fermions) cannot share the same single-particle microstate.

1948   Claude Shannon (1916–2001) initiates the modern study of information theory with the publication of his paper *The Mathematical Theory of Communication*.

1949   Robert Fano (1917–) and, independently, Claude Shannon invent a coding scheme called Fano or Shannon–Fano coding according to which an alphabet of symbols can be efficiently translated into binary digits.

1950   Edward Purcell (1912–1997) and Robert Pound (1919–2010) create a system of nuclear spins that exhibits negative temperature.

1995   Eric Cornell (1961–) and Carl Wieman (1951–) cool a gas of weakly interacting bosons (rubidium atoms) below the critical temperature and produce a Bose–Einstein condensate.

Appendix V

Answers to problems

1.1 The entropy of the fluid system increases because the composite weight–fluid system is an isolated system that suffers an irreversible change.

1.3 $C\ln(T_f/T_i)$.

1.4 $\Delta S = nR\ln(V_f/V_i)$.

1.5 $\Delta S/n = -(R/2)\ln 2$.

1.6 (a) Equations of state (3) and (4) do not observe the first and second laws of thermodynamics. Equations of state (1), (2), and (5) do observe the first and second laws of thermodynamics.

   (b) The entropy function for equations of state (1) is $S(E,V) = a\ln V - b^2/E$, for (2) is $S(E,V) = b\ln(EV)$, and for (5) is $S(E,V) = abV + b\ln E$.

1.7 $T = 1/aV^2$ and $P = 2aEVT$.

1.8 (a) $E = C_V T + (V - V_o)^2 /(2\kappa_{To}V_o)$ and $P = \alpha_{Po}T/\kappa_{To} - (V - V_o)/(\kappa_{To}V_o E_o)$.

   (b) No. As $T \to 0$, S does not go to a finite constant independent of the thermodynamic state variables.

1.9 (1) violates the third law, (2) violates the third law, (3) violates the third law and violates $T > 0$, (4) violates the third law, (5) violates the third law, (6) violates the third law and violates $T > 0$, and (7) violates neither requirement.

2.1 (a) $\Delta S = k\ln(3^{1100}/2^{1000})$.

   (b) $(2/3)^{1000}(1/3)^{100}$.

2.2 (a) 4, (b) 13, (c) 52, and (d) these two events are independent.

2.3 14%, about $n = 90$.

2.4 (a) n^N and 10^{100}.

   (b) $(N+n-1)!/N!(n-1)!$ and 1.34×10^{40}.

   (c) $n!/N!(n-N)!$ and $1.01\cdot10^{29}$.

2.5 (a) $\Omega = N!/[n_+!n_-!]$.

   (b) $S(L) = c + (Nk/2)\{2\ln 2 - (1 - L/Na)\ln(1 - L/Na) - (1 + L/Na)\ln(1 + L/Na)\}$.

   (c) $F = (kT/2a)[\ln(1 - L/Na) - \ln(1 + L/Na)]$.

   (d) $F = -(kT/Na^2)L$.

172

3.1 2.26×10^{19}.

3.2 (a) $E = 3NkT/2 - aN^2/V$.

  (b) $S(E,V,N) = c(N) + Nk\left[\ln(V - Nb) + (3/2)\ln(E + N^2 a/V)\right.$
  $\left. + (3/2)\ln(4\pi e/3Nm)\right]$.

  (d) $P = NkT/(V - Nb) - aN^2/V^2$.

3.3 (a) $E = 5NkT/2$ and $PV = NkT$.

  (b) Each molecule has five degrees of freedom – three translational and two rotational.

3.4 $\Delta S = (2PV/T)\ln 2$.

3.5 $S(E,N) = pN + 3Nk\left[1 + \ln(E/3NH v_o)\right]$ where p is any real number.

3.6 517 m/s^2.

4.1 (a) $h v_o/kT = \ln\left[(x + \frac{1}{2})/(x - \frac{1}{2})\right]$ and $Z_1 = \sqrt{(x - \frac{1}{2})(x + \frac{1}{2})}$.

4.2 6.1×10^{-6}.

4.4 (a) $Z_1 = 1 + e^{-\epsilon/kT}$.

  (b) $E = N\epsilon e^{-\epsilon/kT}/(1 + e^{-\epsilon/kT})$.

  (c) $C = Nk(\epsilon/kT)^2/(1 + e^{\epsilon/kT})^2$.

  (d) When the temperature is sufficiently low the heat capacity is also low because the system cannot absorb any energy unless it absorbs enough to start populating the high energy state. Alternatively, when the temperature is too high, the heat capacity is low because most of the particles are already in the high energy state.

4.7 $\mu/T = 5k/2 - S/N$ where S is given by the Sackur–Tetrode entropy (4.2).

4.8 (a) $n_+/N = e^{2m_B B_o/kT}/(1 + e^{2m_B B_o/kT})$ and $n_-/N = 1/(1 + e^{2m_B B_o/kT})$.

  (b) $\lim_{T \to \infty} n_+/N = 1/2$, $\lim_{T \to 0} n_+/N = 1$, $\lim_{T \to 0} n_-/N = 1/2$, and $\lim_{T \to 0} n_-/N = 0$.

  (c) 0.122 K.

4.9 (a) $C = Nkx^2 e^{2x}/(1 + e^x)^2$ where $x = 2m_B B_0/kT$.

  (b) $C \to 0$ as $T \to \infty$, $T \to -\infty$ and $T \to 0$.

4.10 (a) $M = Nm_B\left[(e^{2m_B B_o/kT} - 1)/(e^{2m_B B_o/kT} + 1)\right]$.

  (c) $C = Nm_B^2/k$.

5.1 (a) Occupied $1/(1 + e^{-\epsilon/kT})$ and unoccupied $e^{-\epsilon/kT}/(1 + e^{-\epsilon/kT})$.

  (b) $S = k\left\{(\epsilon/kT)e^{-\epsilon/kT}/(1 + e^{-\epsilon/kT}) + \ln(1 + e^{-\epsilon/kT})\right\}$.

6.2 (b) $(N/E)^{3/2}(N/V)(h^3/2m)^{2/3} \geq 1$.

  (c) $N/n \geq (3/2\pi e)^{3/2} \approx 0.0742$.

6.4 $NkT/E_o \approx 0.377$. Therefore, the gas is in the quantum regime.

6.5 $R = \left(1/M^{1/3}\right)\left(h^2/Gmm_p^{5/3}\right)\left(5/2\pi e\right)\left(3/4\pi\right)^{2/3}$.

7.1 (a) $P^v = -\left(8\pi kT v^2 \, dv/c^3\right)\ln\left(1-e^{-hv/kT}\right)$.

(b) $P = \left(8\pi^5/45\right)\left(k^4 T^4/h^3 c^3\right)$.

7.2 4.04×10^{13} Pa. The ratio of radiation to gas pressure at the center of the Sun is 1.00×10^{-3}.

7.3 (a) $\varepsilon(\lambda) = \left(8\pi h/c\lambda\right)^5/\left(e^{hc/\lambda kT}-1\right)$.

(b) $\lambda_{max} = 0.2014\left(hc/kT\right)$.

(c) 5800 K.

7.4 $60.4V\left(kT/ch\right)^3$.

7.5 $E = 1.27NkT\left(T/T_c\right)^{3/2}$.

8.2 (a) $2\log_2 4 = \log_2 16$ or 4 bits.

(b) $\log_2 8$ or 3 bits since here the messages rather than the symbols out of which they are composed are equiprobable.

8.3 (a) $4\log_2 27$ or 19.02 bits.

(b) $12\log_2 3$ or 19.02 bits.

8.4 (a) $\log_2 5$ or 2.32 bits.

(b) $\log_2 12$ or 3.58 bits.

8.6 (a) 1.75 bits. (b) $a \leftrightarrow 0$, $b \leftrightarrow 10$, $c \leftrightarrow 110$, and $d \leftrightarrow 111$. (c) 1.75 bits per symbol.

8.7 (a) One set of blanks with X_2 and X_4 and the other with X_3 and X_4. (b) 7 bits and 4 bits.

Appendix VI
Annotated further reading

An item appears here for at least one, and sometimes for both, of two reasons: I consulted it in writing this guide and I recommend it for further study. Publication dates always refer to the first edition, unless two dates appear. Then the earlier date refers to the first edition and the later date to the edition actually consulted.

von Baeyer, Hans Christian. *Maxwell's Demon: Why Warmth Disperses and Time Passes.* 207 pages. Random House, New York, 1998. An engaging, popular presentation of thermodynamics, statistical and quantum mechanics, Maxwell's demon, and information theory.

von Baeyer, Hans Christian. *Information: The New Language of Science.* 272 pages. Harvard University Press, Cambridge, Massachusetts, 2005. Another helpful, popular book by Hans von Baeyer.

Baierlein, Ralph. *Atoms and Information Theory.* 486 pages. W. H. Freeman, San Francisco, California, 1971. A thorough and reliable statistical mechanics text framed in the language of information theory.

Baierlein, Ralph. *Thermal Physics.* 442 pages. Cambridge University Press, Cambridge, UK, 1999. Another quite helpful Baierlein text. The "thermal physics" approach presents thermodynamics and statistical mechanics as one subject.

Boltzmann, Ludwig. *Lectures on Gas Theory.* Translated by Stephen Brush. 490 pages. New York, Dover, 1995 (1896 and 1898). Boltzmann's writing is neither concise nor clear. His derivation of the Maxwell distribution from the hypothesis of equally probable microstates appears on pages 55–58.

Carnot, Sadi and others. *Reflections on the Motive Power of Fire And Other Papers on the Second Law of Thermodynamics.* Dover, New York, 2005. Contains the title essay as well as historically important essays by Emile Clapeyron and Rudolf Clausius.

Cover, Thomas M. and Joy Thomas. *Elements of Information Theory, 2nd edition.* 776 pages. Wiley, Hoboken, New Jersey, 2006. A standard information theory text with a theorem-proof presentation of the subject. Aimed at upper level undergraduate and graduate students.

Feynman, Richard P., Robert B. Leighton and Mathew Sands. *The Feynman Lectures on Physics*, Vol. 3. Addison-Wesley, Reading, Massachusetts, 1965. Section 3–4 has a nice discussion of the way the superposition of the wave functions of identical particles generates the distinction between bosons and fermions.

Gibbs, J. Willard. *Elementary Principles in Statistical Mechanics*. 207 pages. Ox Bow Press, Woodbridge, Connecticut, 1981 (1902). A masterful if difficult presentation of classical statistical mechanics by one of its originators.

Glieck, James. *The Information*. 526 pages. Pantheon, New York, 2011. An expansive, popular history of information from cuneiform to cyberspace.

Gould, Harvey and Jan Tobochnik. *Statistical and Thermal Physics*. 511 pages. Princeton, Princeton, New Jersey, 2010. A thermal physics text with fully integrated computational examples and exercises.

Jackson, E. Atlee. *Equilibrium Statistical Mechanics*. 242 pages. Dover Publications, New York, 2000 (1968). A compact introduction to statistical mechanics. Jackson's short appendix "Concerning the Entropy Constant" is noteworthy.

Jaynes, E. T. *Information theory and statistical mechanics*. Physical Review, Volume 166, pp. 620–630, May 15, 1957. Jaynes makes a strong case for the "maximum-entropy" information-theoretic approach to statistical mechanics.

Hartley, R. V. L. Transmission of information. *Bell System Technical Journal*, Volume 7, Number 3, pp. 535–563, July 1928. A precursor of Shannon's 1948 paper on information theory.

Kuhn, Thomas. *Black-Body Theory and the Quantum Discontinuity 1894–1912*. 378 pages. University of Chicago Press, Chicago, Illinois, 1978. History of science at its best. Kuhn argues that in 1900 Planck did not realize he had quantized anything. In the eight years following 1900 others, notably Albert Einstein, forced Planck to recognize the revolutionary nature of his own work.

Leff, Harvey and Andrew F. Rex. *Maxwell's Demon 2: Entropy, Classical and Quantum Information, Computing*. 485 pages. Institute of Physics, Bristol, U.K., 2003. An anthology of papers inspired by Maxwell's imagined entropy-reducing beast that search out possible connections, formal and substantial, between the entropy of physics and chemistry and the entropy of information theory. Complete with an introductory essay and a collection of demon visuals.

Lemons, Don S. *Mere Thermodynamics*. 207 pages. Johns Hopkins, Baltimore, Maryland, 2008. A presentation of thermodynamic laws, methods, and applications that emphasizes the logical structure of thermodynamics.

Lindley, David. *Boltzmann's Atom*. 260 pages. Simon & Schuster, New York, 2001. A delightful non-technical biography of Boltzmann that skillfully weaves together the content and context of Boltzmann's science.

Pais, Abraham. *Subtle is the Lord* .... 552 pages. Oxford University Press, Oxford, UK, 1982. A biography of Einstein that contains detailed explanations of his physics and the physics of his era. Some formal background in physics is required to fully appreciate this biography.

Penrose, Oliver. *Foundations of Statistical Mechanics: A Deductive Treatment*. 272 pages. Dover, New York, 2005 (1970). Quite valuable but at a considerably higher level and density of mathematics than this guide.

Planck, Max. *The Theory of Heat Radiation*. 224 pages. Dover, New York, 1991 (1914). Planck absorbed Boltzmann's ideas in order to articulate the statistical mechanics of blackbody radiation. Admirably clear if also tedious in detail and somewhat dated. Influenced the presentation of ideas in Chapters 2 and 4 of this Guide.

Schroeder, Daniel V. *Thermal Physics*. 422 pages. Addison Wesley, Upper Saddle, New Jersey, 1999. A deservedly popular thermal physics textbook.

Shannon, Claude and Warren Weaver. *The Mathematical Theory of Communication*. 125 pages. University of Illinois Press, Urbana, Illinois, 1964 (1949). Contains the 1948 paper that initiated information theory. Shannon's paper is fairly mathematical relative to most of the items in this bibliography. However, Shannon also motivates and explains. Weaver's short expository article on information theory originally appeared in *Scientific American*.

Singh, Jagjit. *Great Ideas in Information Theory, Language and Cybernetics*. 338 pages. Dover, New York, 1966. The first seven chapters (83 pages) are a nice non-technical introduction to information theory.

Index

Printed in the United States
By Bookmasters